Monte Carlo Methods

A Hands-On Computational Introduction
Utilizing Excel

Synthesis Lectures on Mathematics and Statistics

Editor
Steven G. Krantz, *Washington University, St. Louis*

iv

The Geometry of Walker Manifolds
Miguel Brozos-Vázquez, Eduardo García-Río, Peter Gilkey, Stana Nikčević, and Ramón Vázquez-Lorenzo
2009

An Introduction to Multivariable Mathematics
Leon Simon
2008

Jordan Canonical Form: Application to Differential Equations
Steven H. Weintraub
2008

Statistics is Easy!
Dennis Shasha and Manda Wilson
2008

A Gyrovector Space Approach to Hyperbolic Geometry
Abraham Albert Ungar
2008

Monte Carlo Methods: A Hands-On Computational Introduction Utilizing Excel
Sujaul Chowdhury

ISBN: 978-3-031-01301-0   paperback
ISBN: 978-3-031-02429-0   ebook
ISBN: 978-3-031-00275-5   hardcover

DOI 10.1007/978-3-031-02429-0

A Publication in the Springer series
SYNTHESIS LECTURES ON MATHEMATICS AND STATISTICS

Lecture #37
Series Editor: Steven G. Krantz, *Washington University, St. Louis*
Series ISSN
Print 1938-1743   Electronic 1938-1751

Monte Carlo Methods

A Hands-On Computational Introduction Utilizing Excel

Sujaul Chowdhury
Shahjalal University of Science and Technology

SYNTHESIS LECTURES ON MATHEMATICS AND STATISTICS #37

ABSTRACT

This book is intended for undergraduate students of Mathematics, Statistics, and Physics who know nothing about Monte Carlo Methods but wish to know how they work. All treatments have been done as much manually as is practicable. The treatments are deliberately manual to let the readers get the real feel of how Monte Carlo Methods work.

Definite integrals of a total of five functions $F(x)$, namely $\operatorname{Sin}(x)$, $\operatorname{Cos}(x)$, e^x, $\log_e(x)$, and $1/(1 + x^2)$, have been evaluated using constant, linear, Gaussian, and exponential probability density functions $p(x)$. It is shown that results agree with known exact values better if $p(x)$ is proportional to $F(x)$. Deviation from the proportionality results in worse agreement.

This book is on Monte Carlo Methods which are numerical methods for Computational Physics. These are parts of a syllabus for undergraduate students of Mathematics and Physics for the course titled "Computational Physics."

Need for the book: Besides the three referenced books, this is the only book that teaches how basic Monte Carlo methods work. This book is much more explicit and easier to follow than the three referenced books. The two chapters on the Variational Quantum Monte Carlo method are additional contributions of the book.

Pedagogical features: After a thorough acquaintance with background knowledge in Chapter 1, five *thoroughly worked out* examples on how to carry out Monte Carlo integration is included in Chapter 2. Moreover, the book contains two chapters on the Variational Quantum Monte Carlo method applied to a simple harmonic oscillator and a hydrogen atom.

The book is a good read; it is intended to make readers adept at using the method. The book is intended to aid in hands-on learning of the Monte Carlo methods.

KEYWORDS

Monte Carlo methods, basic Monte Carlo integration, variational quantum Monte Carlo method, simple harmonic oscillator, hydrogen atom

Contents

CHAPTER 1

Introduction

1.1 RANDOM VARIABLE

In layman terms, a random variable refers to a variable the value of which is not predictable. We do not know what value the variable can or will assume. In connection with the Monte Carlo method, the term "random variable" has a precise meaning. We do not know the value of the variable in any given case, but we do know the values that the variable can assume and the probabilities of these values. The result of a single trial cannot be precisely predicted, but the result of a large number of trials can be predicted very reliably. To define a random variable, we must indicate the values that the variable can assume and the probabilities of occurrence of these values.

1.2 CONTINUOUS RANDOM VARIABLE

A random variable x is called "continuous" if it can assume any (fractional) value in a certain interval, say a to b. Besides specifying the interval containing all its possible values, we need to state a function $p(x)$ called probability density function or probability distribution function.

We have:

1. $p(x) \geq 0$ for $a \leq x \leq b$,

2. the product $p(x)dx$ is probability that $x \leq x \leq x + dx$,

3. $\int_{a'}^{b'} p(x)dx$ is probability that $a' \leq x \leq b'$ where $a < a'$ and $b' < b$, and

4. $\int_{a}^{b} p(x)dx = 1$ indicating the surety that x lies in the interval a to b.

As such, the average or so-called expectation value of x is given by $A_x = \int_{a}^{b} x p(x)dx$, and average value of any function of x, say $f(x)$, is $A_f = \int_{a}^{b} f(x)p(x)dx$.

1.3 UNIFORM RANDOM VARIABLE

A random variable u in the interval $0 \leq x \leq 1$ having constant, say C, probability density function $p(x) = C$ is said to be a uniform random variable. The requirement that $\int_{0}^{1} p(x)dx = 1$ gives $\int_{0}^{1} C dx = 1$ which gives $C = 1$. Thus, probability density function of uniform random variable is $p(x) = 1$.

Here is a manual way of obtaining values of a uniform random variable. Take ten table tennis balls; write on each of them one of the ten digits 0, 1, 2, 3, …, 9; mingle them well in a box; take one ball out and note the digit; place the ball back to the box and mingle the ten balls well again; then take one ball out and note the digit. Continue the process until we get a table of 300 digits. From this table, take a sequence of three digits, say 1, 6, 9; construct the number 169; then divide the number by 1000 and get 0.169. Repeat the process until we get a list of 100 numbers such as 0.169. Thus, we have a table of 100 numbers in the interval 0 to 1. Since each of the digits 0, 1, 2, 3, …, 9 was equally likely to be picked up during the trials, the said 100 numbers are uniformly distributed in the interval 0 to 1. Thus, we have 100 values of uniform random variable u. See Table 1.1.

1.4   NORMAL OR GAUSSIAN RANDOM VARIABLE

Probability density function $p(x)$ for normal random variable x is given by

$$p(x) = \frac{1}{\sigma\sqrt{2\pi}}\mathrm{Exp}\left[-\frac{1}{2}\left(\frac{x-a}{\sigma}\right)^2\right], \tag{1.1}$$

defined for $-\infty < x < +\infty$. The average value of x is a and variance of x is σ^2. According to the so-called rule of 3 sigma, $\int_{a-3\sigma}^{a+3\sigma} p(x)dx \approx 0.997$.

1.5   TRANSFORMATION OR MODELING OF RANDOM VARIABLE

Values of random variable η corresponding to a given probability density function $p(x)$ defined for $a \leq x \leq b$ can be obtained by transformation of values of one "standard" random variable which is usually taken as the uniform random variable u (distributed in the interval $0 \leq x \leq 1$). See Table 1.1.

We already know the values of u. The process of finding the values of η by transforming the values of u is called modeling or transformation of the random variable. With every available value of u, the corresponding value of η is obtainable by solving the equation

$$\int_a^\eta p(x)dx = u. \tag{1.2}$$

This is verified below. Here values of η are distributed in the interval $a \leq x \leq b$ with probability density function $p(x)$.

Let

$$y(x) = \int_a^x p(x)dx. \tag{1.3}$$

Table 1.1: Showing 100 uniform random numbers u_i (in the interval 0–1)

i	u_i	i	u_i	i	u_i	i	u_i
1	0.169	26	0.225	51	0.393	76	0.231
2	0.050	27	0.687	52	0.306	77	0.836
3	0.844	28	0.060	53	0.819	78	0.167
4	0.182	29	0.070	54	0.620	79	0.545
5	0.599	30	0.835	55	0.875	80	0.748
6	0.985	31	0.466	56	0.773	81	0.440
7	0.726	32	0.271	57	0.492	82	0.674
8	0.148	33	0.813	58	0.332	83	0.287
9	0.788	34	0.376	59	0.185	84	0.052
10	0.698	35	0.983	60	0.565	85	0.940
11	0.010	36	0.181	61	0.434	86	0.679
12	0.397	37	0.583	62	0.665	87	0.910
13	0.852	38	0.612	63	0.516	88	0.254
14	0.761	39	0.468	64	0.659	89	0.236
15	0.309	40	0.614	65	0.709	90	0.854
16	0.113	41	0.768	66	0.515	91	0.877
17	0.212	42	0.811	67	0.683	92	0.483
18	0.809	43	0.461	68	0.092	93	0.117
19	0.084	44	0.985	69	0.417	94	0.847
20	0.276	45	0.735	70	0.598	95	0.201
21	0.243	46	0.559	71	0.682	96	0.294
22	0.989	47	0.884	72	0.948	97	0.071
23	0.508	48	0.229	73	0.125	98	0.065
24	0.817	49	0.920	74	0.149	99	0.740
25	0.174	50	0.524	75	0.246	100	0.954

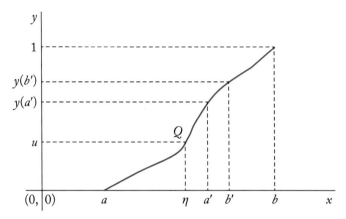

Figure 1.1: Figure to help explain transformation or modeling of random variable.

Since $p(x) > 0$, and $\int_a^b p(x)dx = 1$, we have

$$y(a) = \int_a^a p(x)dx = 0$$

$$y(b) = \int_a^b p(x)dx = 1$$

emphasizing that $p(x)$ is normalized, and

$$y'(x) = p(x) > 0.$$

This means that y monotonically increases from 0 to 1 for the interval $a < x < b$. See Figure 1.1. Straight line $y = u$ where $0 < u < 1$ intersects the curve $y(x)$ at Q giving us the value of η which satisfies Equation (1.2): $\int_a^\eta p(x)dx = u$.

$$p(a' < \eta < b') = p(y(a') < u < y(b')) = y(b') - y(a')$$

$$= \int_a^{b'} p(x)dx - \int_a^{a'} p(x)dx = \int_{a'}^{b'} p(x)dx.$$

Thus, random variable η which obeys Equation (1.2): $\int_a^\eta p(x)dx = u$ has the probability density function $p(x)$.

1.6 EXAMPLES OF TRANSFORMATION OR MODELING OF RANDOM VARIABLE

Values of random variable η obeying normalized, arbitrary probability density function $p(x)$ in the interval $a < x < b$ are obtainable from Equation (1.2): $\int_a^\eta p(x)dx = u$ where u is uniform random variable in the interval 0 to 1.

Example 1.1 Values of random variable η uniformly distributed in the interval $a < x < b$ are obtainable from $\int_a^\eta C \, dx = u$ where $p(x)$ should be normalized, i.e., $\int_a^b p(x)dx = 1$, $\int_a^b C dx = 1$, $C = 1/(b-a)$. Thus, $\int_a^\eta \frac{1}{b-a} dx = u$. Hence,

$$\eta = a + (b-a)u. \tag{1.4}$$

This is a linear transformation of u. Use of tabulated values of u (see Table 1.1) in Equation (1.4) gives us values of random variable η obeying uniform probability density function $p(x) = 1/(b-a)$ in the interval $a < x < b$. See Table 1.2 which shows 100 uniform random numbers η_i in the interval $-\pi/2$ to $+\pi/2$ obtained using Equation (1.4): $\eta = a + (b-a)u$ which gives η_i's as a linear transformation of u_i's.

Example 1.2 We now discuss modeling of random variable g obeying Gaussian or normal probability density function given by

$$p_g(x; \, a = 0, \, \sigma = 1) = \frac{1}{\sigma\sqrt{2\pi}}\text{Exp}\left[-\frac{1}{2}\left(\frac{x-a}{\sigma}\right)^2\right]. \tag{1.5}$$

Values of g are obtainable from Equation (1.2): $\int_a^\eta p(x)dx = u$ with p given by Equation (1.5) with $a = 0$ and $\sigma = 1$, i.e., from $\int_{a-3\sigma}^g \frac{1}{\sigma\sqrt{2\pi}}\text{Exp}\left[-\frac{1}{2}\left(\frac{x-a}{\sigma}\right)^2\right] dx = u$ with $a = 0$ and $\sigma = 1$, i.e., from

$$\int_{-3}^g \frac{1}{\sqrt{2\pi}}\text{Exp}\left[-\frac{1}{2}x^2\right] dx = u. \tag{1.6}$$

But Equation (1.6) is *not* analytically solvable for g. As such, we have used Program 1.3 in Mathematica 6.0 and obtained values of $\int_{-3}^\eta \frac{1}{\sqrt{2\pi}}\text{Exp}\left[-\frac{1}{2}x^2\right] dx$ for η ranging from -3 to $+3$ in step of 0.005. According to Equation (1.6), values of g are those values of η for which $\int_{-3}^\eta \frac{1}{\sqrt{2\pi}}\text{Exp}\left[-\frac{1}{2}x^2\right] dx = u$. Hence, we have *literally* and *manually* hunted for each value of u of Table 1.1 in the 3rd and 6th columns of Table 1.3 and wrote down the corresponding value of η from the 2nd and 5th columns of Table 1.3 in Table 1.4.

Table 1.2: Showing 100 uniform random numbers η_i in the interval $-\pi/2$ to $+\pi/2$ obtained using Equation (1.4): $\eta = a + (b - a)u$ which gives η_i's as linear transformation of u_i's

i	u_i	$\eta_i =$ $-\pi/2 + \pi u_i$	i	u_i	$\eta_i =$ $-\pi/2 + \pi u_i$	i	u_i	$\eta_i =$ $-\pi/2 + \pi u_i$	i	u_i	$\eta_i =$ $-\pi/2 + \pi u_i$
1	0.169	−1.040	26	0.225	−0.864	51	0.393	−0.336	76	0.231	−0.845
2	0.050	−1.414	27	0.687	0.587	52	0.306	−0.609	77	0.836	1.056
3	0.844	1.081	28	0.060	−1.382	53	0.819	1.002	78	0.167	−1.046
4	0.182	−0.999	29	0.070	−1.351	54	0.620	0.377	79	0.545	0.141
5	0.599	0.311	30	0.835	1.052	55	0.875	1.178	80	0.748	0.779
6	0.985	1.524	31	0.466	−0.107	56	0.773	0.858	81	0.440	−0.188
7	0.726	0.710	32	0.271	−0.719	57	0.492	−0.025	82	0.674	0.547
8	0.148	−1.106	33	0.813	0.983	58	0.332	−0.528	83	0.287	−0.669
9	0.788	0.905	34	0.376	−0.390	59	0.185	−0.990	84	0.052	−1.407
10	0.698	0.622	35	0.983	1.517	60	0.565	0.204	85	0.940	1.382
11	0.010	−1.539	36	0.181	−1.002	61	0.434	−0.207	86	0.679	0.562
12	0.397	−0.324	37	0.583	0.261	62	0.665	0.518	87	0.910	1.288
13	0.852	1.106	38	0.612	0.352	63	0.516	0.050	88	0.254	−0.773
14	0.761	0.820	39	0.468	−0.101	64	0.659	0.500	89	0.236	−0.829
15	0.309	−0.600	40	0.614	0.358	65	0.709	0.657	90	0.854	1.112
16	0.113	−1.216	41	0.768	0.842	66	0.515	0.047	91	0.877	1.184
17	0.212	−0.905	42	0.811	0.977	67	0.683	0.575	92	0.483	−0.053
18	0.809	0.971	43	0.461	−0.123	68	0.092	−1.282	93	0.117	−1.203
19	0.084	−1.307	44	0.985	1.524	69	0.417	−0.261	94	0.847	1.090
20	0.276	−0.704	45	0.735	0.738	70	0.598	0.308	95	0.201	−0.939
21	0.243	−0.807	46	0.559	0.185	71	0.682	0.572	96	0.294	−0.647
22	0.989	1.536	47	0.884	1.206	72	0.948	1.407	97	0.071	−1.348
23	0.508	0.025	48	0.229	−0.851	73	0.125	−1.178	98	0.065	−1.367
24	0.817	0.996	49	0.920	1.319	74	0.149	−1.103	99	0.740	0.754
25	0.174	−1.024	50	0.524	0.075	75	0.246	−0.798	100	0.954	1.426

Table 1.3: Showing values of $\int_{-3}^{\eta} \frac{1}{\sqrt{2\pi}} \mathrm{Exp}\left[-\frac{1}{2}x^2\right] dx$ for η ranging from -3 to $+3$ in step of 0.005. Tabulated using Program 1.3 via Microsoft Excel.

i	η	$\int_{-3}^{\eta} \frac{1}{\sqrt{2\pi}} \exp\left[-\frac{1}{2}x^2\right] dx$	i	η	$\int_{-3}^{\eta} \frac{1}{\sqrt{2\pi}} \exp\left[-\frac{1}{2}x^2\right] dx$
1	−3.000	0.000	26	−2.875	0.001
2	−2.995	0.000	27	−2.870	0.001
3	−2.990	0.000	28	−2.865	0.001
4	−2.985	0.000	29	−2.860	0.001
5	−2.980	0.000	30	−2.855	0.001
6	−2.975	0.000	31	−2.850	0.001
7	−2.970	0.000	32	−2.845	0.001
8	−2.965	0.000	33	−2.840	0.001
9	−2.960	0.000	34	−2.835	0.001
10	−2.955	0.000	35	−2.830	0.001
11	−2.950	0.000	36	−2.825	0.001
12	−2.945	0.000	37	−2.820	0.001
13	−2.940	0.000	38	−2.815	0.001
14	−2.935	0.000	39	−2.810	0.001
15	−2.930	0.000	40	−2.805	0.001
16	−2.925	0.000	41	−2.800	0.001
17	−2.920	0.000	42	−2.795	0.001
18	−2.915	0.000	43	−2.790	0.001
19	−2.910	0.000	44	−2.785	0.001
20	−2.905	0.000	45	−2.780	0.001
21	−2.900	0.001	46	−2.775	0.001
22	−2.895	0.001	47	−2.770	0.001
23	−2.890	0.001	48	−2.765	0.001
24	−2.885	0.001	49	−2.760	0.002
25	−2.880	0.001	50	−2.755	0.002

i	η	$\int_{-3}^{\eta}\frac{1}{\sqrt{2\pi}}\exp\left[-\frac{1}{2}x^2\right]dx$	i	η	$\int_{-3}^{\eta}\frac{1}{\sqrt{2\pi}}\exp\left[-\frac{1}{2}x^2\right]dx$
51	−2.750	0.002	76	−2.625	0.003
52	−2.745	0.002	77	−2.620	0.003
53	−2.740	0.002	78	−2.615	0.003
54	−2.735	0.002	79	−2.610	0.003
55	−2.730	0.002	80	−2.605	0.003
56	−2.725	0.002	81	−2.600	0.003
57	−2.720	0.002	82	−2.595	0.003
58	−2.715	0.002	83	−2.590	0.003
59	−2.710	0.002	84	−2.585	0.004
60	−2.705	0.002	85	−2.580	0.004
61	−2.700	0.002	86	−2.575	0.004
62	−2.695	0.002	87	−2.570	0.004
63	−2.690	0.002	88	−2.565	0.004
64	−2.685	0.002	89	−2.560	0.004
65	−2.680	0.002	90	−2.555	0.004
66	−2.675	0.002	91	−2.550	0.004
67	−2.670	0.002	92	−2.545	0.004
68	−2.665	0.002	93	−2.540	0.004
69	−2.660	0.003	94	−2.535	0.004
70	−2.655	0.003	95	−2.530	0.004
71	−2.650	0.003	96	−2.525	0.004
72	−2.645	0.003	97	−2.520	0.005
73	−2.640	0.003	98	−2.515	0.005
74	−2.635	0.003	99	−2.510	0.005
75	−2.630	0.003	100	−2.505	0.005

i	η	$\int_{-3}^{\eta} \frac{1}{\sqrt{2\pi}} \exp\left[-\frac{1}{2}x^2\right] dx$	i	η	$\int_{-3}^{\eta} \frac{1}{\sqrt{2\pi}} \exp\left[-\frac{1}{2}x^2\right] dx$
101	−2.500	0.005	126	−2.375	0.007
102	−2.495	0.005	127	−2.370	0.008
103	−2.490	0.005	128	−2.365	0.008
104	−2.485	0.005	129	−2.360	0.008
105	−2.480	0.005	130	−2.355	0.008
106	−2.475	0.005	131	−2.350	0.008
107	−2.470	0.005	132	−2.345	0.008
108	−2.465	0.006	133	−2.340	0.008
109	−2.460	0.006	134	−2.335	0.008
110	−2.455	0.006	135	−2.330	0.009
111	−2.450	0.006	136	−2.325	0.009
112	−2.445	0.006	137	−2.320	0.009
113	−2.440	0.006	138	−2.315	0.009
114	−2.435	0.006	139	−2.310	0.009
115	−2.430	0.006	140	−2.305	0.009
116	−2.425	0.006	141	−2.300	0.009
117	−2.420	0.006	142	−2.295	0.010
118	−2.415	0.007	143	−2.290	0.010
119	−2.410	0.007	144	−2.285	0.010
120	−2.405	0.007	145	−2.280	0.010
121	−2.400	0.007	146	−2.275	0.010
122	−2.395	0.007	147	−2.270	0.010
123	−2.390	0.007	148	−2.265	0.010
124	−2.385	0.007	149	−2.260	0.011
125	−2.380	0.007	150	−2.255	0.011

i	η	$\int\limits_{-3}^{\eta}\frac{1}{\sqrt{2\pi}}\exp\left[-\frac{1}{2}x^2\right]dx$	i	η	$\int\limits_{-3}^{\eta}\frac{1}{\sqrt{2\pi}}\exp\left[-\frac{1}{2}x^2\right]dx$
151	−2.250	0.011	176	−2.125	0.015
152	−2.245	0.011	177	−2.120	0.016
153	−2.240	0.011	178	−2.115	0.016
154	−2.235	0.011	179	−2.110	0.016
155	−2.230	0.012	180	−2.105	0.016
156	−2.225	0.012	181	−2.100	0.017
157	−2.220	0.012	182	−2.095	0.017
158	−2.215	0.012	183	−2.090	0.017
159	−2.210	0.012	184	−2.085	0.017
160	−2.205	0.012	185	−2.080	0.017
161	−2.200	0.013	186	−2.075	0.018
162	−2.195	0.013	187	−2.070	0.018
163	−2.190	0.013	188	−2.065	0.018
164	−2.185	0.013	189	−2.060	0.018
165	−2.180	0.013	190	−2.055	0.019
166	−2.175	0.013	191	−2.050	0.019
167	−2.170	0.014	192	−2.045	0.019
168	−2.165	0.014	193	−2.040	0.019
169	−2.160	0.014	194	−2.035	0.020
170	−2.155	0.014	195	−2.030	0.020
171	−2.150	0.014	196	−2.025	0.020
172	−2.145	0.015	197	−2.020	0.020
173	−2.140	0.015	198	−2.015	0.021
174	−2.135	0.015	199	−2.010	0.021
175	−2.130	0.015	200	−2.005	0.021

i	η	$\int\limits_{-3}^{\eta}\frac{1}{\sqrt{2\pi}}\exp\left[-\frac{1}{2}x^2\right]dx$	i	η	$\int\limits_{-3}^{\eta}\frac{1}{\sqrt{2\pi}}\exp\left[-\frac{1}{2}x^2\right]dx$
201	−2.000	0.021	226	−1.875	0.029
202	−1.995	0.022	227	−1.870	0.029
203	−1.990	0.022	228	−1.865	0.030
204	−1.985	0.022	229	−1.860	0.030
205	−1.980	0.023	230	−1.855	0.030
206	−1.975	0.023	231	−1.850	0.031
207	−1.970	0.023	232	−1.845	0.031
208	−1.965	0.023	233	−1.840	0.032
209	−1.960	0.024	234	−1.835	0.032
210	−1.955	0.024	235	−1.830	0.032
211	−1.950	0.024	236	−1.825	0.033
212	−1.945	0.025	237	−1.820	0.033
213	−1.940	0.025	238	−1.815	0.033
214	−1.935	0.025	239	−1.810	0.034
215	−1.930	0.025	240	−1.805	0.034
216	−1.925	0.026	241	−1.800	0.035
217	−1.920	0.026	242	−1.795	0.035
218	−1.915	0.026	243	−1.790	0.035
219	−1.910	0.027	244	−1.785	0.036
220	−1.905	0.027	245	−1.780	0.036
221	−1.900	0.027	246	−1.775	0.037
222	−1.895	0.028	247	−1.770	0.037
223	−1.890	0.028	248	−1.765	0.037
224	−1.885	0.028	249	−1.760	0.038
225	−1.880	0.029	250	−1.755	0.038

i	η	$\int\limits_{-3}^{\eta} \frac{1}{\sqrt{2\pi}} \exp\left[-\frac{1}{2}x^2\right] dx$	i	η	$\int\limits_{-3}^{\eta} \frac{1}{\sqrt{2\pi}} \exp\left[-\frac{1}{2}x^2\right] dx$
251	−1.750	0.039	276	−1.625	0.051
252	−1.745	0.039	277	−1.620	0.051
253	−1.740	0.040	278	−1.615	0.052
254	−1.735	0.040	279	−1.610	0.052
255	−1.730	0.040	280	−1.605	0.053
256	−1.725	0.041	281	−1.600	0.053
257	−1.720	0.041	282	−1.595	0.054
258	−1.715	0.042	283	−1.590	0.055
259	−1.710	0.042	284	−1.585	0.055
260	−1.705	0.043	285	−1.580	0.056
261	−1.700	0.043	286	−1.575	0.056
262	−1.695	0.044	287	−1.570	0.057
263	−1.690	0.044	288	−1.565	0.057
264	−1.685	0.045	289	−1.560	0.058
265	−1.680	0.045	290	−1.555	0.059
266	−1.675	0.046	291	−1.550	0.059
267	−1.670	0.046	292	−1.545	0.060
268	−1.665	0.047	293	−1.540	0.060
269	−1.660	0.047	294	−1.535	0.061
270	−1.655	0.048	295	−1.530	0.062
271	−1.650	0.048	296	−1.525	0.062
272	−1.645	0.049	297	−1.520	0.063
273	−1.640	0.049	298	−1.515	0.064
274	−1.635	0.050	299	−1.510	0.064
275	−1.630	0.050	300	−1.505	0.065

i	η	$\int_{-3}^{\eta} \frac{1}{\sqrt{2\pi}} \exp\left[-\frac{1}{2}x^2\right]dx$	i	η	$\int_{-3}^{\eta} \frac{1}{\sqrt{2\pi}} \exp\left[-\frac{1}{2}x^2\right]dx$
301	−1.500	0.065	326	−1.375	0.083
302	−1.495	0.066	327	−1.370	0.084
303	−1.490	0.067	328	−1.365	0.085
304	−1.485	0.067	329	−1.360	0.086
305	−1.480	0.068	330	−1.355	0.086
306	−1.475	0.069	331	−1.350	0.087
307	−1.470	0.069	332	−1.345	0.088
308	−1.465	0.070	333	−1.340	0.089
309	−1.460	0.071	334	−1.335	0.090
310	−1.455	0.071	335	−1.330	0.090
311	−1.450	0.072	336	−1.325	0.091
312	−1.445	0.073	337	−1.320	0.092
313	−1.440	0.074	338	−1.315	0.093
314	−1.435	0.074	339	−1.310	0.094
315	−1.430	0.075	340	−1.305	0.095
316	−1.425	0.076	341	−1.300	0.095
317	−1.420	0.076	342	−1.295	0.096
318	−1.415	0.077	343	−1.290	0.097
319	−1.410	0.078	344	−1.285	0.098
320	−1.405	0.079	345	−1.280	0.099
321	−1.400	0.079	346	−1.275	0.100
322	−1.395	0.080	347	−1.270	0.101
323	−1.390	0.081	348	−1.265	0.102
324	−1.385	0.082	349	−1.260	0.102
325	−1.380	0.082	350	−1.255	0.103

i	η	$\int\limits_{-3}^{\eta}\frac{1}{\sqrt{2\pi}}\exp\left[-\frac{1}{2}x^2\right]dx$	i	η	$\int\limits_{-3}^{\eta}\frac{1}{\sqrt{2\pi}}\exp\left[-\frac{1}{2}x^2\right]dx$
351	−1.250	0.104	376	−1.125	0.129
352	−1.245	0.105	377	−1.120	0.130
353	−1.240	0.106	378	−1.115	0.131
354	−1.235	0.107	379	−1.110	0.132
355	−1.230	0.108	380	−1.105	0.133
356	−1.225	0.109	381	−1.100	0.134
357	−1.220	0.110	382	−1.095	0.135
358	−1.215	0.111	383	−1.090	0.137
359	−1.210	0.112	384	−1.085	0.138
360	−1.205	0.113	385	−1.080	0.139
361	−1.200	0.114	386	−1.075	0.140
362	−1.195	0.115	387	−1.070	0.141
363	−1.190	0.116	388	−1.065	0.142
364	−1.185	0.117	389	−1.060	0.143
365	−1.180	0.118	390	−1.055	0.144
366	−1.175	0.119	391	−1.050	0.146
367	−1.170	0.120	392	−1.045	0.147
368	−1.165	0.121	393	−1.040	0.148
369	−1.160	0.122	394	−1.035	0.149
370	−1.155	0.123	395	−1.030	0.150
371	−1.150	0.124	396	−1.025	0.151
372	−1.145	0.125	397	−1.020	0.153
373	−1.140	0.126	398	−1.015	0.154
374	−1.135	0.127	399	−1.010	0.155
375	−1.130	0.128	400	−1.005	0.156

i	η	$\int_{-3}^{\eta} \frac{1}{\sqrt{2\pi}} \exp\left[-\frac{1}{2}x^2\right] dx$	i	η	$\int_{-3}^{\eta} \frac{1}{\sqrt{2\pi}} \exp\left[-\frac{1}{2}x^2\right] dx$
401	−1.000	0.157	426	−0.875	0.189
402	−0.995	0.159	427	−0.870	0.191
403	−0.990	0.160	428	−0.865	0.192
404	−0.985	0.161	429	−0.860	0.194
405	−0.980	0.162	430	−0.855	0.195
406	−0.975	0.163	431	−0.850	0.196
407	−0.970	0.165	432	−0.845	0.198
408	−0.965	0.166	433	−0.840	0.199
409	−0.960	0.167	434	−0.835	0.201
410	−0.955	0.168	435	−0.830	0.202
411	−0.950	0.170	436	−0.825	0.203
412	−0.945	0.171	437	−0.820	0.205
413	−0.940	0.172	438	−0.815	0.206
414	−0.935	0.174	439	−0.810	0.208
415	−0.930	0.175	440	−0.805	0.209
416	−0.925	0.176	441	−0.800	0.211
417	−0.920	0.177	442	−0.795	0.212
418	−0.915	0.179	443	−0.790	0.213
419	−0.910	0.180	444	−0.785	0.215
420	−0.905	0.181	445	−0.780	0.216
421	−0.900	0.183	446	−0.775	0.218
422	−0.895	0.184	447	−0.770	0.219
423	−0.890	0.185	448	−0.765	0.221
424	−0.885	0.187	449	−0.760	0.222
425	−0.880	0.188	450	−0.755	0.224

i	η	$\int\limits_{-3}^{\eta} \frac{1}{\sqrt{2\pi}}\exp\left[-\frac{1}{2}x^2\right]dx$	i	η	$\int\limits_{-3}^{\eta} \frac{1}{\sqrt{2\pi}}\exp\left[-\frac{1}{2}x^2\right]dx$
451	−0.750	0.225	476	−0.625	0.265
452	−0.745	0.227	477	−0.620	0.266
453	−0.740	0.228	478	−0.615	0.268
454	−0.735	0.230	479	−0.610	0.270
455	−0.730	0.231	480	−0.605	0.271
456	−0.725	0.233	481	−0.600	0.273
457	−0.720	0.234	482	−0.595	0.275
458	−0.715	0.236	483	−0.590	0.276
459	−0.710	0.238	484	−0.585	0.278
460	−0.705	0.239	485	−0.580	0.280
461	−0.700	0.241	486	−0.575	0.281
462	−0.695	0.242	487	−0.570	0.283
463	−0.690	0.244	488	−0.565	0.285
464	−0.685	0.245	489	−0.560	0.286
465	−0.680	0.247	490	−0.555	0.288
466	−0.675	0.248	491	−0.550	0.290
467	−0.670	0.250	492	−0.545	0.292
468	−0.665	0.252	493	−0.540	0.293
469	−0.660	0.253	494	−0.535	0.295
470	−0.655	0.255	495	−0.530	0.297
471	−0.650	0.256	496	−0.525	0.298
472	−0.645	0.258	497	−0.520	0.300
473	−0.640	0.260	498	−0.515	0.302
474	−0.635	0.261	499	−0.510	0.304
475	−0.630	0.263	500	−0.505	0.305

i	η	$\int\limits_{-3}^{\eta} \frac{1}{\sqrt{2\pi}}\exp\left[-\frac{1}{2}x^2\right]dx$	i	η	$\int\limits_{-3}^{\eta} \frac{1}{\sqrt{2\pi}}\exp\left[-\frac{1}{2}x^2\right]dx$
501	−0.500	0.307	526	−0.375	0.352
502	−0.495	0.309	527	−0.370	0.354
503	−0.490	0.311	528	−0.365	0.356
504	−0.485	0.312	529	−0.360	0.358
505	−0.480	0.314	530	−0.355	0.360
506	−0.475	0.316	531	−0.350	0.362
507	−0.470	0.318	532	−0.345	0.364
508	−0.465	0.320	533	−0.340	0.366
509	−0.460	0.321	534	−0.335	0.367
510	−0.455	0.323	535	−0.330	0.369
511	−0.450	0.325	536	−0.325	0.371
512	−0.445	0.327	537	−0.320	0.373
513	−0.440	0.329	538	−0.315	0.375
514	−0.435	0.330	539	−0.310	0.377
515	−0.430	0.332	540	−0.305	0.379
516	−0.425	0.334	541	−0.300	0.381
517	−0.420	0.336	542	−0.295	0.383
518	−0.415	0.338	543	−0.290	0.385
519	−0.410	0.340	544	−0.285	0.386
520	−0.405	0.341	545	−0.280	0.388
521	−0.400	0.343	546	−0.275	0.390
522	−0.395	0.345	547	−0.270	0.392
523	−0.390	0.347	548	−0.265	0.394
524	−0.385	0.349	549	−0.260	0.396
525	−0.380	0.351	550	−0.255	0.398

i	η	$\int\limits_{-3}^{\eta} \frac{1}{\sqrt{2\pi}} \exp\left[-\frac{1}{2}x^2\right] dx$	i	η	$\int\limits_{-3}^{\eta} \frac{1}{\sqrt{2\pi}} \exp\left[-\frac{1}{2}x^2\right] dx$
551	−0.250	0.400	576	−0.125	0.449
552	−0.245	0.402	577	−0.120	0.451
553	−0.240	0.404	578	−0.115	0.453
554	−0.235	0.406	579	−0.110	0.455
555	−0.230	0.408	580	−0.105	0.457
556	−0.225	0.410	581	−0.100	0.459
557	−0.220	0.412	582	−0.095	0.461
558	−0.215	0.414	583	−0.090	0.463
559	−0.210	0.415	584	−0.085	0.465
560	−0.205	0.417	585	−0.080	0.467
561	−0.200	0.419	586	−0.075	0.469
562	−0.195	0.421	587	−0.070	0.471
563	−0.190	0.423	588	−0.065	0.473
564	−0.185	0.425	589	−0.060	0.475
565	−0.180	0.427	590	−0.055	0.477
566	−0.175	0.429	591	−0.050	0.479
567	−0.170	0.431	592	−0.045	0.481
568	−0.165	0.433	593	−0.040	0.483
569	−0.160	0.435	594	−0.035	0.485
570	−0.155	0.437	595	−0.030	0.487
571	−0.150	0.439	596	−0.025	0.489
572	−0.145	0.441	597	−0.020	0.491
573	−0.140	0.443	598	−0.015	0.493
574	−0.135	0.445	599	−0.010	0.495
575	−0.130	0.447	600	−0.005	0.497

i	η	$\int_{-3}^{\eta} \frac{1}{\sqrt{2\pi}}\exp\left[-\frac{1}{2}x^2\right]dx$	i	η	$\int_{-3}^{\eta} \frac{1}{\sqrt{2\pi}}\exp\left[-\frac{1}{2}x^2\right]dx$
601	0.000	0.499	626	0.125	0.548
602	0.005	0.501	627	0.130	0.550
603	0.010	0.503	628	0.135	0.552
604	0.015	0.505	629	0.140	0.554
605	0.020	0.507	630	0.145	0.556
606	0.025	0.509	631	0.150	0.558
607	0.030	0.511	632	0.155	0.560
608	0.035	0.513	633	0.160	0.562
609	0.040	0.515	634	0.165	0.564
610	0.045	0.517	635	0.170	0.566
611	0.050	0.519	636	0.175	0.568
612	0.055	0.521	637	0.180	0.570
613	0.060	0.523	638	0.185	0.572
614	0.065	0.525	639	0.190	0.574
615	0.070	0.527	640	0.195	0.576
616	0.075	0.529	641	0.200	0.578
617	0.080	0.531	642	0.205	0.580
618	0.085	0.533	643	0.210	0.582
619	0.090	0.535	644	0.215	0.584
620	0.095	0.536	645	0.220	0.586
621	0.100	0.538	646	0.225	0.588
622	0.105	0.540	647	0.230	0.590
623	0.110	0.542	648	0.235	0.592
624	0.115	0.544	649	0.240	0.593
625	0.120	0.546	650	0.245	0.595

i	η	$\int\limits_{-3}^{\eta} \frac{1}{\sqrt{2\pi}}\exp\left[-\frac{1}{2}x^2\right]dx$	i	η	$\int\limits_{-3}^{\eta} \frac{1}{\sqrt{2\pi}}\exp\left[-\frac{1}{2}x^2\right]dx$
651	0.250	0.597	676	0.375	0.645
652	0.255	0.599	677	0.380	0.647
653	0.260	0.601	678	0.385	0.649
654	0.265	0.603	679	0.390	0.650
655	0.270	0.605	680	0.395	0.652
656	0.275	0.607	681	0.400	0.654
657	0.280	0.609	682	0.405	0.656
658	0.285	0.611	683	0.410	0.658
659	0.290	0.613	684	0.415	0.660
660	0.295	0.615	685	0.420	0.661
661	0.300	0.617	686	0.425	0.663
662	0.305	0.618	687	0.430	0.665
663	0.310	0.620	688	0.435	0.667
664	0.315	0.622	689	0.440	0.669
665	0.320	0.624	690	0.445	0.670
666	0.325	0.626	691	0.450	0.672
667	0.330	0.628	692	0.455	0.674
668	0.335	0.630	693	0.460	0.676
669	0.340	0.632	694	0.465	0.678
670	0.345	0.634	695	0.470	0.679
671	0.350	0.635	696	0.475	0.681
672	0.355	0.637	697	0.480	0.683
673	0.360	0.639	698	0.485	0.685
674	0.365	0.641	699	0.490	0.687
675	0.370	0.643	700	0.495	0.688

i	η	$\int_{-3}^{\eta} \frac{1}{\sqrt{2\pi}} \exp\left[-\frac{1}{2}x^2\right] dx$	i	η	$\int_{-3}^{\eta} \frac{1}{\sqrt{2\pi}} \exp\left[-\frac{1}{2}x^2\right] dx$
701	0.500	0.690	726	0.625	0.733
702	0.505	0.692	727	0.630	0.734
703	0.510	0.694	728	0.635	0.736
704	0.515	0.695	729	0.640	0.738
705	0.520	0.697	730	0.645	0.739
706	0.525	0.699	731	0.650	0.741
707	0.530	0.701	732	0.655	0.742
708	0.535	0.702	733	0.660	0.744
709	0.540	0.704	734	0.665	0.746
710	0.545	0.706	735	0.670	0.747
711	0.550	0.707	736	0.675	0.749
712	0.555	0.709	737	0.680	0.750
713	0.560	0.711	738	0.685	0.752
714	0.565	0.713	739	0.690	0.754
715	0.570	0.714	740	0.695	0.755
716	0.575	0.716	741	0.700	0.757
717	0.580	0.718	742	0.705	0.758
718	0.585	0.719	743	0.710	0.760
719	0.590	0.721	744	0.715	0.761
720	0.595	0.723	745	0.720	0.763
721	0.600	0.724	746	0.725	0.764
722	0.605	0.726	747	0.730	0.766
723	0.610	0.728	748	0.735	0.767
724	0.615	0.729	749	0.740	0.769
725	0.620	0.731	750	0.745	0.771

i	η	$\int\limits_{-3}^{\eta}\frac{1}{\sqrt{2\pi}}\exp\left[-\frac{1}{2}x^2\right]dx$	i	η	$\int\limits_{-3}^{\eta}\frac{1}{\sqrt{2\pi}}\exp\left[-\frac{1}{2}x^2\right]dx$
751	0.750	0.772	776	0.875	0.808
752	0.755	0.774	777	0.880	0.809
753	0.760	0.775	778	0.885	0.811
754	0.765	0.777	779	0.890	0.812
755	0.770	0.778	780	0.895	0.813
756	0.775	0.779	781	0.900	0.815
757	0.780	0.781	782	0.905	0.816
758	0.785	0.782	783	0.910	0.817
759	0.790	0.784	784	0.915	0.819
760	0.795	0.785	785	0.920	0.820
761	0.800	0.787	786	0.925	0.821
762	0.805	0.788	787	0.930	0.822
763	0.810	0.790	788	0.935	0.824
764	0.815	0.791	789	0.940	0.825
765	0.820	0.793	790	0.945	0.826
766	0.825	0.794	791	0.950	0.828
767	0.830	0.795	792	0.955	0.829
768	0.835	0.797	793	0.960	0.830
769	0.840	0.798	794	0.965	0.831
770	0.845	0.800	795	0.970	0.833
771	0.850	0.801	796	0.975	0.834
772	0.855	0.802	797	0.980	0.835
773	0.860	0.804	798	0.985	0.836
774	0.865	0.805	799	0.990	0.838
775	0.870	0.806	800	0.995	0.839

i	η	$\int_{-3}^{\eta} \frac{1}{\sqrt{2\pi}}\exp\left[-\frac{1}{2}x^2\right]dx$	i	η	$\int_{-3}^{\eta} \frac{1}{\sqrt{2\pi}}\exp\left[-\frac{1}{2}x^2\right]dx$
801	1.000	0.840	826	1.125	0.868
802	1.005	0.841	827	1.130	0.869
803	1.010	0.842	828	1.135	0.870
804	1.015	0.844	829	1.140	0.872
805	1.020	0.845	830	1.145	0.873
806	1.025	0.846	831	1.150	0.874
807	1.030	0.847	832	1.155	0.875
808	1.035	0.848	833	1.160	0.876
809	1.040	0.849	834	1.165	0.877
810	1.045	0.851	835	1.170	0.878
811	1.050	0.852	836	1.175	0.879
812	1.055	0.853	837	1.180	0.880
813	1.060	0.854	838	1.185	0.881
814	1.065	0.855	839	1.190	0.882
815	1.070	0.856	840	1.195	0.883
816	1.075	0.857	841	1.200	0.884
817	1.080	0.859	842	1.205	0.885
818	1.085	0.860	843	1.210	0.886
819	1.090	0.861	844	1.215	0.886
820	1.095	0.862	845	1.220	0.887
821	1.100	0.863	846	1.225	0.888
822	1.105	0.864	847	1.230	0.889
823	1.110	0.865	848	1.235	0.890
824	1.115	0.866	849	1.240	0.891
825	1.120	0.867	850	1.245	0.892

i	η	$\int\limits_{-3}^{\eta}\dfrac{1}{\sqrt{2\pi}}\exp\left[-\dfrac{1}{2}x^2\right]dx$	i	η	$\int\limits_{-3}^{\eta}\dfrac{1}{\sqrt{2\pi}}\exp\left[-\dfrac{1}{2}x^2\right]dx$
851	1.250	0.893	876	1.375	0.914
852	1.255	0.894	877	1.380	0.915
853	1.260	0.895	878	1.385	0.916
854	1.265	0.896	879	1.390	0.916
855	1.270	0.897	880	1.395	0.917
856	1.275	0.897	881	1.400	0.918
857	1.280	0.898	882	1.405	0.919
858	1.285	0.899	883	1.410	0.919
859	1.290	0.900	884	1.415	0.920
860	1.295	0.901	885	1.420	0.921
861	1.300	0.902	886	1.425	0.922
862	1.305	0.903	887	1.430	0.922
863	1.310	0.904	888	1.435	0.923
864	1.315	0.904	889	1.440	0.924
865	1.320	0.905	890	1.445	0.924
866	1.325	0.906	891	1.450	0.925
867	1.330	0.907	892	1.455	0.926
868	1.335	0.908	893	1.460	0.927
869	1.340	0.909	894	1.465	0.927
870	1.345	0.909	895	1.470	0.928
871	1.350	0.910	896	1.475	0.929
872	1.355	0.911	897	1.480	0.929
873	1.360	0.912	898	1.485	0.930
874	1.365	0.913	899	1.490	0.931
875	1.370	0.913	900	1.495	0.931

i	η	$\int_{-3}^{\eta} \frac{1}{\sqrt{2\pi}} \exp\left[-\frac{1}{2}x^2\right] dx$	i	η	$\int_{-3}^{\eta} \frac{1}{\sqrt{2\pi}} \exp\left[-\frac{1}{2}x^2\right] dx$
901	1.500	0.932	926	1.625	0.947
902	1.505	0.932	927	1.630	0.947
903	1.510	0.933	928	1.635	0.948
904	1.515	0.934	929	1.640	0.948
905	1.520	0.934	930	1.645	0.949
906	1.525	0.935	931	1.650	0.949
907	1.530	0.936	932	1.655	0.950
908	1.535	0.936	933	1.660	0.950
909	1.540	0.937	934	1.665	0.951
910	1.545	0.937	935	1.670	0.951
911	1.550	0.938	936	1.675	0.952
912	1.555	0.939	937	1.680	0.952
913	1.560	0.939	938	1.685	0.953
914	1.565	0.940	939	1.690	0.953
915	1.570	0.940	940	1.695	0.954
916	1.575	0.941	941	1.700	0.954
917	1.580	0.942	942	1.705	0.955
918	1.585	0.942	943	1.710	0.955
919	1.590	0.943	944	1.715	0.955
920	1.595	0.943	945	1.720	0.956
921	1.600	0.944	946	1.725	0.956
922	1.605	0.944	947	1.730	0.957
923	1.610	0.945	948	1.735	0.957
924	1.615	0.945	949	1.740	0.958
925	1.620	0.946	950	1.745	0.958

i	η	$\int\limits_{-3}^{\eta} \frac{1}{\sqrt{2\pi}}\exp\left[-\frac{1}{2}x^2\right]dx$	i	η	$\int\limits_{-3}^{\eta} \frac{1}{\sqrt{2\pi}}\exp\left[-\frac{1}{2}x^2\right]dx$
951	1.750	0.959	976	1.875	0.968
952	1.755	0.959	977	1.880	0.969
953	1.760	0.959	978	1.885	0.969
954	1.765	0.960	979	1.890	0.969
955	1.770	0.960	980	1.895	0.970
956	1.775	0.961	981	1.900	0.970
957	1.780	0.961	982	1.905	0.970
958	1.785	0.962	983	1.910	0.971
959	1.790	0.962	984	1.915	0.971
960	1.795	0.962	985	1.920	0.971
961	1.800	0.963	986	1.925	0.972
962	1.805	0.963	987	1.930	0.972
963	1.810	0.964	988	1.935	0.972
964	1.815	0.964	989	1.940	0.972
965	1.820	0.964	990	1.945	0.973
966	1.825	0.965	991	1.950	0.973
967	1.830	0.965	992	1.955	0.973
968	1.835	0.965	993	1.960	0.974
969	1.840	0.966	994	1.965	0.974
970	1.845	0.966	995	1.970	0.974
971	1.850	0.966	996	1.975	0.975
972	1.855	0.967	997	1.980	0.975
973	1.860	0.967	998	1.985	0.975
974	1.865	0.968	999	1.990	0.975
975	1.870	0.968	1000	1.995	0.976

i	η	$\int_{-3}^{\eta} \frac{1}{\sqrt{2\pi}} \exp\left[-\frac{1}{2}x^2\right] dx$	i	η	$\int_{-3}^{\eta} \frac{1}{\sqrt{2\pi}} \exp\left[-\frac{1}{2}x^2\right] dx$
1001	2.000	0.976	1026	2.125	0.982
1002	2.005	0.976	1027	2.130	0.982
1003	2.010	0.976	1028	2.135	0.982
1004	2.015	0.977	1029	2.140	0.982
1005	2.020	0.977	1030	2.145	0.983
1006	2.025	0.977	1031	2.150	0.983
1007	2.030	0.977	1032	2.155	0.983
1008	2.035	0.978	1033	2.160	0.983
1009	2.040	0.978	1034	2.165	0.983
1010	2.045	0.978	1035	2.170	0.984
1011	2.050	0.978	1036	2.175	0.984
1012	2.055	0.979	1037	2.180	0.984
1013	2.060	0.979	1038	2.185	0.984
1014	2.065	0.979	1039	2.190	0.984
1015	2.070	0.979	1040	2.195	0.985
1016	2.075	0.980	1041	2.200	0.985
1017	2.080	0.980	1042	2.205	0.985
1018	2.085	0.980	1043	2.210	0.985
1019	2.090	0.980	1044	2.215	0.985
1020	2.095	0.981	1045	2.220	0.985
1021	2.100	0.981	1046	2.225	0.986
1022	2.105	0.981	1047	2.230	0.986
1023	2.110	0.981	1048	2.235	0.986
1024	2.115	0.981	1049	2.240	0.986
1025	2.120	0.982	1050	2.245	0.986

i	η	$\int_{-3}^{\eta} \frac{1}{\sqrt{2\pi}} \exp\left[-\frac{1}{2}x^2\right] dx$	i	η	$\int_{-3}^{\eta} \frac{1}{\sqrt{2\pi}} \exp\left[-\frac{1}{2}x^2\right] dx$
1051	2.250	0.986	1076	2.375	0.990
1052	2.255	0.987	1077	2.380	0.990
1053	2.260	0.987	1078	2.385	0.990
1054	2.265	0.987	1079	2.390	0.990
1055	2.270	0.987	1080	2.395	0.990
1056	2.275	0.987	1081	2.400	0.990
1057	2.280	0.987	1082	2.405	0.991
1058	2.285	0.987	1083	2.410	0.991
1059	2.290	0.988	1084	2.415	0.991
1060	2.295	0.988	1085	2.420	0.991
1061	2.300	0.988	1086	2.425	0.991
1062	2.305	0.988	1087	2.430	0.991
1063	2.310	0.988	1088	2.435	0.991
1064	2.315	0.988	1089	2.440	0.991
1065	2.320	0.988	1090	2.445	0.991
1066	2.325	0.989	1091	2.450	0.992
1067	2.330	0.989	1092	2.455	0.992
1068	2.335	0.989	1093	2.460	0.992
1069	2.340	0.989	1094	2.465	0.992
1070	2.345	0.989	1095	2.470	0.992
1071	2.350	0.989	1096	2.475	0.992
1072	2.355	0.989	1097	2.480	0.992
1073	2.360	0.990	1098	2.485	0.992
1074	2.365	0.990	1099	2.490	0.992
1075	2.370	0.990	1100	2.495	0.992

i	η	$\int_{-3}^{\eta} \frac{1}{\sqrt{2\pi}} \exp\left[-\frac{1}{2}x^2\right] dx$	i	η	$\int_{-3}^{\eta} \frac{1}{\sqrt{2\pi}} \exp\left[-\frac{1}{2}x^2\right] dx$
1101	2.500	0.992	1126	2.625	0.994
1102	2.505	0.993	1127	2.630	0.994
1103	2.510	0.993	1128	2.635	0.994
1104	2.515	0.993	1129	2.640	0.995
1105	2.520	0.993	1130	2.645	0.995
1106	2.525	0.993	1131	2.650	0.995
1107	2.530	0.993	1132	2.655	0.995
1108	2.535	0.993	1133	2.660	0.995
1109	2.540	0.993	1134	2.665	0.995
1110	2.545	0.993	1135	2.670	0.995
1111	2.550	0.993	1136	2.675	0.995
1112	2.555	0.993	1137	2.680	0.995
1113	2.560	0.993	1138	2.685	0.995
1114	2.565	0.993	1139	2.690	0.995
1115	2.570	0.994	1140	2.695	0.995
1116	2.575	0.994	1141	2.700	0.995
1117	2.580	0.994	1142	2.705	0.995
1118	2.585	0.994	1143	2.710	0.995
1119	2.590	0.994	1144	2.715	0.995
1120	2.595	0.994	1145	2.720	0.995
1121	2.600	0.994	1146	2.725	0.995
1122	2.605	0.994	1147	2.730	0.995
1123	2.610	0.994	1148	2.735	0.996
1124	2.615	0.994	1149	2.740	0.996
1125	2.620	0.994	1150	2.745	0.996

i	η	$\int_{-3}^{\eta} \frac{1}{\sqrt{2\pi}} \exp\left[-\frac{1}{2}x^2\right] dx$	i	η	$\int_{-3}^{\eta} \frac{1}{\sqrt{2\pi}} \exp\left[-\frac{1}{2}x^2\right] dx$
1151	2.750	0.996	1176	2.875	0.997
1152	2.755	0.996	1177	2.880	0.997
1153	2.760	0.996	1178	2.885	0.997
1154	2.765	0.996	1179	2.890	0.997
1155	2.770	0.996	1180	2.895	0.997
1156	2.775	0.996	1181	2.900	0.997
1157	2.780	0.996	1182	2.905	0.997
1158	2.785	0.996	1183	2.910	0.997
1159	2.790	0.996	1184	2.915	0.997
1160	2.795	0.996	1185	2.920	0.997
1161	2.800	0.996	1186	2.925	0.997
1162	2.805	0.996	1187	2.930	0.997
1163	2.810	0.996	1188	2.935	0.997
1164	2.815	0.996	1189	2.940	0.997
1165	2.820	0.996	1190	2.945	0.997
1166	2.825	0.996	1191	2.950	0.997
1167	2.830	0.996	1192	2.955	0.997
1168	2.835	0.996	1193	2.960	0.997
1169	2.840	0.996	1194	2.965	0.997
1170	2.845	0.996	1195	2.970	0.997
1171	2.850	0.996	1196	2.975	0.997
1172	2.855	0.996	1197	2.980	0.997
1173	2.860	0.997	1198	2.985	0.997
1174	2.865	0.997	1199	2.990	0.997
1175	2.870	0.997	1200	2.995	0.997
			1201	3.000	0.997

Table 1.4: Showing 100 random numbers g obeying Gaussian or Normal probability density function $p_g(x) = \frac{1}{\sigma\sqrt{2\pi}}\text{Exp}\left[-\frac{1}{2}\left(\frac{x-a}{\sigma}\right)^2\right]$ with $a = 0$ and $\sigma = 1$. u_i's are corresponding values of uniform random numbers in the interval 0 to 1 (of Table 1.1).

i	u_i	g_i	i	u_i	g_i	i	u_i	g_i	i	u_i	g_i
1	0.169	−0.955	26	0.225	−0.750	51	0.393	−0.270	76	0.231	−0.730
2	0.050	−1.635	27	0.687	0.490	52	0.306	−0.505	77	0.836	0.985
3	0.844	1.015	28	0.060	−1.540	53	0.819	0.915	78	0.167	−0.960
4	0.182	−0.905	29	0.070	−1.465	54	0.620	0.310	79	0.545	0.115
5	0.599	0.255	30	0.835	0.980	55	0.875	1.155	80	0.748	0.670
6	0.985	2.195	31	0.466	−0.085	56	0.773	0.750	81	0.440	−0.150
7	0.726	0.605	32	0.271	−0.605	57	0.492	−0.020	82	0.674	0.455
8	0.148	−1.040	33	0.813	0.895	58	0.332	−0.430	83	0.287	−0.560
9	0.788	0.805	34	0.376	−0.315	59	0.185	−0.890	84	0.052	−1.615
10	0.698	0.520	35	0.983	2.145	60	0.565	0.165	85	0.940	1.565
11	0.010	−2.295	36	0.181	−0.905	61	0.434	−0.165	86	0.679	0.470
12	0.397	−0.255	37	0.583	0.210	62	0.665	0.430	87	0.910	1.350
13	0.852	1.050	38	0.612	0.285	63	0.516	0.040	88	0.254	−0.660
14	0.761	0.715	39	0.468	−0.080	64	0.659	0.410	89	0.236	−0.715
15	0.309	−0.495	40	0.614	0.290	65	0.709	0.555	90	0.854	1.060
16	0.113	−1.205	41	0.768	0.735	66	0.515	0.040	91	0.877	1.165
17	0.212	−0.795	42	0.811	0.885	67	0.683	0.480	92	0.483	−0.040
18	0.809	0.880	43	0.461	−0.095	68	0.092	−1.320	93	0.117	−1.185
19	0.084	−1.370	44	0.985	2.195	69	0.417	−0.205	94	0.847	1.030
20	0.276	−0.590	45	0.735	0.630	70	0.598	0.250	95	0.201	−0.835
21	0.243	−0.695	46	0.559	0.150	71	0.682	0.475	96	0.294	−0.540
22	0.989	2.325	47	0.884	1.200	72	0.948	1.635	97	0.071	−1.460
23	0.508	0.020	48	0.229	−0.740	73	0.125	−1.145	98	0.065	−1.505
24	0.817	0.910	49	0.920	1.415	74	0.149	−1.035	99	0.740	0.645
25	0.174	−0.935	50	0.524	0.060	75	0.246	−0.685	100	0.954	1.695

Program 1.3

```
h=0.005;
eta=-3.005;
i=0;
Table[{i=i+1,eta=eta+h,int=(1/Sqrt[2*3.1416])*Integrate
        [Exp[-0.5*x^2], {x,-3,eta}]},{eta,-3.005,2.995,h}];
TableForm[%,TableSpacing->{2,2},
        TableHeadings->{None,{"i","eta","integral"}}]
```

Example 1.4   We now discuss modeling of random variable G obeying Gaussian or Normal probability density function given by

$$P_G(x) = \frac{1}{\sigma\sqrt{2\pi}}\text{Exp}\left[-\frac{1}{2}\left(\frac{x-a}{\sigma}\right)^2\right] \qquad (1.7)$$

with non-zero value of a and $\sigma \neq 1$. It is well known that values of G are related to those of g via

$$G = a + \sigma g, \qquad (1.8)$$

where values of g are random numbers obeying Equation (1.7) with $a = 0$ and $\sigma = 1$.

1.7   VARIANCE REDUCTION AND IMPORTANCE SAMPLING

Let us consider a definite integral

$$I = \int_a^b F(x)dx. \qquad (1.9)$$

Let us re-write Equation (1.9) as

$$I = \int_a^b \frac{F(x)}{p(x)}p(x)dx = \int_a^b f(x)p(x)dx \qquad (1.10)$$

where $f(x) = F(x)/p(x)$.

As discussed in Section 1.2, average value of any function of x, say $f(x)$, is given by $A_f = \int_a^b f(x)p(x)dx$ where $p(x)$ is a *normalized* probability density function. Thus, according to Equation (1.10), value I of the integral is average value of F/p.

Table 1.5: Showing 100 random numbers G obeying Gaussian or Normal probability density function $P_G(x) = \frac{1}{\sigma\sqrt{2\pi}}\text{Exp}\left[-\frac{1}{2}\left(\frac{x-a}{\sigma}\right)^2\right]$ with $a = \pi/2$ and $\sigma = 0.5$. u_i's are corresponding values of uniform random numbers in the interval 0 to 1 and g's are corresponding values of random numbers obeying $p_g(x) = \frac{1}{\sigma\sqrt{2\pi}}\text{Exp}\left[-\frac{1}{2}\left(\frac{x-a}{\sigma}\right)^2\right]$ with $a = 0$ and $\sigma = 1$. Values of G have been obtained using $G = a + \sigma g = \pi/2 + 0.5g$.

i	u_i	g_i	G_i	i	u_i	g_i	G_i
1	0.169	−0.955	1.093	26	0.225	−0.750	1.196
2	0.050	−1.635	0.753	27	0.687	0.490	1.816
3	0.844	1.015	2.078	28	0.060	−1.540	0.801
4	0.182	−0.905	1.118	29	0.070	−1.465	0.838
5	0.599	0.255	1.698	30	0.835	0.980	2.061
6	0.985	2.195	2.668	31	0.466	−0.085	1.528
7	0.726	0.605	1.873	32	0.271	−0.605	1.268
8	0.148	−1.040	1.051	33	0.813	0.895	2.018
9	0.788	0.805	1.973	34	0.376	−0.315	1.413
10	0.698	0.520	1.831	35	0.983	2.145	2.643
11	0.010	−2.295	0.423	36	0.181	−0.905	1.118
12	0.397	−0.255	1.443	37	0.583	0.210	1.676
13	0.852	1.050	2.096	38	0.612	0.285	1.713
14	0.761	0.715	1.928	39	0.468	−0.080	1.531
15	0.309	−0.495	1.323	40	0.614	0.290	1.716
16	0.113	−1.205	0.968	41	0.768	0.735	1.938
17	0.212	−0.795	1.173	42	0.811	0.885	2.013
18	0.809	0.880	2.011	43	0.461	−0.095	1.523
19	0.084	−1.370	0.886	44	0.985	2.195	2.668
20	0.276	−0.590	1.276	45	0.735	0.630	1.886
21	0.243	−0.695	1.223	46	0.559	0.150	1.646
22	0.989	2.325	2.733	47	0.884	1.200	2.171
23	0.508	0.020	1.581	48	0.229	−0.740	1.201
24	0.817	0.910	2.026	49	0.920	1.415	2.278
25	0.174	−0.935	1.103	50	0.524	0.060	1.601

i	u_i	g_i	G_i	i	u_i	g_i	G_i
51	0.393	−0.270	1.436	76	0.231	−0.730	1.206
52	0.306	−0.505	1.318	77	0.836	0.985	2.063
53	0.819	0.915	2.028	78	0.167	−0.960	1.091
54	0.620	0.310	1.726	79	0.545	0.115	1.628
55	0.875	1.155	2.148	80	0.748	0.670	1.906
56	0.773	0.750	1.946	81	0.440	−0.150	1.496
57	0.492	−0.020	1.561	82	0.674	0.455	1.798
58	0.332	−0.430	1.356	83	0.287	−0.560	1.291
59	0.185	−0.890	1.126	84	0.052	−1.615	0.763
60	0.565	0.165	1.653	85	0.940	1.565	2.353
61	0.434	−0.165	1.488	86	0.679	0.470	1.806
62	0.665	0.430	1.786	87	0.910	1.350	2.246
63	0.516	0.040	1.591	88	0.254	−0.660	1.241
64	0.659	0.410	1.776	89	0.236	−0.715	1.213
65	0.709	0.555	1.848	90	0.854	1.060	2.101
66	0.515	0.040	1.591	91	0.877	1.165	2.153
67	0.683	0.480	1.811	92	0.483	−0.040	1.551
68	0.092	−1.320	0.911	93	0.117	−1.185	0.978
69	0.417	−0.205	1.468	94	0.847	1.030	2.086
70	0.598	0.250	1.696	95	0.201	−0.835	1.153
71	0.682	0.475	1.808	96	0.294	−0.540	1.301
72	0.948	1.635	2.388	97	0.071	−1.460	0.841
73	0.125	−1.145	0.998	98	0.065	−1.505	0.818
74	0.149	−1.035	1.053	99	0.740	0.645	1.893
75	0.246	−0.685	1.228	100	0.954	1.695	2.418

We can choose any functional form for $p(x)$, but as proved below, choosing $p(x)$ as proportional to $F(x)$ ensures that the variance of $f(x) = F/p$ is small. This will ensure a better value of the average and hence a better value of the integral $I = \int_a^b F(x)dx$.

Variance of $f(x)$ is square of standard deviation. Variance of $f(x)$ is given by

$$V[f(x)] = A\left[(f - A_f)^2\right],\tag{1.11}$$

where A stands for average value. Thus,

$$Vf(x) = A\left[f^2 - 2fA_f + A_f^2\right] = A\left(f^2\right) - 2A_f A_f + A_f^2$$
$$= A(f^2) - A_f^2$$

or,

$$Vf(x) = \int_a^b f^2(x)p(x)dx - \left(\int_a^b f(x)p(x)dx\right)^2$$
$$= \int_a^b \frac{F^2}{p^2}pdx - \left(\int_a^b \frac{F}{p}p(x)dx\right)^2 = \int_a^b \frac{F^2}{p}dx - \left(\int_a^b Fdx\right)^2.$$

Thus, the variance is

$$Vf(x) = \int_a^b \frac{F^2}{p}dx - I^2.\tag{1.12}$$

If $p(x)$ is taken as

$$p(x) = \frac{|F(x)|}{\int_a^b |F(x)|\,dx}\tag{1.13}$$

the variance given by Equation (1.12) becomes

$$Vf(x) = \int_a^b \frac{F^2}{\dfrac{|F(x)|}{\int_a^b |F(x)|\,dx}}dx - I^2$$

or,

$$Vf(x) = \left(\int_a^b |F(x)|\,dx\right)\left(\int_a^b |F(x)|\,dx\right) - I^2 = \left(\int_a^b |F(x)|\,dx\right)^2 - I^2.\tag{1.14}$$

Equation (1.14) in conjunction with Equation (1.9): $I = \int_a^b F(x)dx$ reveals that the variance vanishes if the integrand $F(x)$ does not change sign. If it does, the variance will be small if condition laid by Equation (1.13) is met. Thus, the probability density function $p(x)$ should be proportional to the integrand $|F(x)|$. This is the so-called *importance sampling*.

Looking back at Equation (1.10), we find that value of the integral is approximately determined by the value of the average

$$I = \frac{1}{N} \sum_{i=1}^{N} \frac{F(x_i)}{p(x_i)}.$$

As recommended by importance sampling, we need to take $p(x) = C \ F(x)$ where C is a constant; we need to normalize $p(x)$ first; $\int_a^b p(x)dx = 1$ gives $\int_a^b C \ F(x)dx = 1$ or,

$$C = \frac{1}{\int_a^b F(x)dx}.$$

Thus,

$$p(x) = \frac{F(x)}{\int_a^b F(x)dx}.$$

The sum

$$\frac{1}{N} \sum_{i=1}^{N} \frac{F(x_i)}{p(x_i)}$$

becomes

$$\frac{1}{N} \sum_{i=1}^{N} \frac{F(x_i)}{\frac{F(x_i)}{\int_a^b F(x)dx}} = \int_a^b F(x)dx.$$

Thus, the sum

$$\frac{1}{N} \sum_{i=1}^{N} \frac{F(x_i)}{p(x_i)}$$

equals the integral if probability density function $p(x)$ is taken proportional to the integrand $|F(x)|$. Slight variation of $p(x)$ from proportionality with $|F(x)|$ will result in slight difference of

$$\frac{1}{N} \sum_{i=1}^{N} \frac{F(x_i)}{p(x_i)}$$

from the actual value of the integral $I = \int_a^b F(x)dx$.

CHAPTER 2

Evaluation of Definite Integrals Using the Monte Carlo Method

2.1 EVALUATION OF DEFINITE INTEGRALS USING THE MONTE CARLO METHOD: EXAMPLE I

As Example I, we now take up the integral

$$I = \int_a^b F(x)dx = \int_0^\pi \text{Sin}(x)dx, \tag{2.1}$$

where $F(x) = \text{Sin}(x)$. We re-write Equation (2.1) as

$$I = \int_a^b \frac{F(x)}{p(x)} p(x)dx \tag{2.2}$$

which, as discussed in Section 1.7, implies that average value of F/p is the value of the integral, i.e.,

$$I = \frac{1}{N} \sum_{i=1}^N \frac{F(x_i)}{p(x_i)}. \tag{2.3}$$

where x_i's are random values of x in the interval $a < x < b$ obeying probability density function $p(x)$.

Here $p(x)$ is a suitable probability density function which, as demonstrated in Section 1.7, at least should follow $F(x)$, if not be proportional to $F(x)$. See Figure 2.1 in which we have plotted $F(x) = \text{Sin}(x)$ along with normalized probability density function of a uniform random variable in the interval 0 to π given by $p(x) = 1/\pi$. Let $p(x) = C$, a constant. Normalization requires that $\int_0^\pi Cdx = 1$ which gives $C = 1/\pi$. Uniform $p(x) = 1/\pi$ is not too different from a slowly varying function of x like $\text{Sin}(x)$. Hence, $p(x) = 1/\pi$ is a good choice.

We now generate random variable η obeying $p(x) = 1/\pi$ in the interval $0 < x < \pi$ using Equation (1.4): $\eta = a + (b - a)u$ where u is random variable (of Table 1.1) of uniform probability density function in the interval $0 < x < 1$. Thus, we have $\eta = \pi u$. That $\eta = \pi u$ can be calculated using $\int_a^\eta p(x)dx = u$ which gives $\int_0^\eta (1/\pi)dx = u$ or, $\eta = \pi u$.

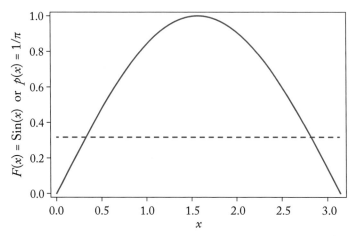

Figure 2.1: Showing $F(x) = \mathrm{Sin}(x)$ as undashed curve and $p(x) = 1/\pi$ as dashed curve obtained using the command
```
Plot[{Sin[x],1/Pi},{x,0,Pi},Frame->True,FrameLabel->
{"x","F(x) = Sin(x)  or  p(x) = 1/\[Pi]"},PlotStyle->
{{Black},{Dashed,Black}}]
```
in Mathematica 6.0.

Using Table 2.1 in Microsoft Excel, we have evaluated the integral with the result 1.973 rather than 2. The difference is attributable to the fact that $F(x)$ and $p(x)$ used are not proportional to each other.

We now take up a linear variation for $p(x)$ given by $p(x) = Cx$ where C is a constant. Normalization requires that $\int_0^\pi Cx\,dx = 1$ which gives $C = 2/\pi^2$. Thus, normalized $p(x) = 2x/\pi^2$. See Figure 2.2 in which we have plotted $F(x) = \mathrm{Sin}(x)$ along with the normalized linear probability density function $p(x) = 2x/\pi^2$. The linear probability density function $p(x) = 2x/\pi^2$ is not too different from a slowly varying function of x like $\mathrm{Sin}(x)$. Hence, it is a good choice.

We now generate random variable η obeying $p(x) = 2x/\pi^2$ in the interval $0 < x < \pi$ using Equation (1.2): $\int_a^\eta p(x)dx = u$ or, $\int_0^\eta p(x)dx = u$ where u is random variable (of Table 1.1) of uniform probability density function in the interval $0 < x < 1$. Thus, we have $\eta = \pi u^{1/2}$.

Using Table 2.2 in Microsoft Excel, we have evaluated the integral with the result 1.947 rather than 2. The difference is attributable to the fact that $F(x)$ and $p(x)$ used are not proportional to each other.

We now take up a normalized *Gaussian* variation for $p(x)$ given by $p(x) = \frac{1}{\sigma\sqrt{2\pi}}\mathrm{Exp}\left[-\frac{1}{2}\left(\frac{x-a}{\sigma}\right)^2\right]$ where a is average and σ is variance of x. See Figure 2.3 in which we have plotted $F(x) = \mathrm{Sin}(x)$ along with the normalized Gaussian probability density function

Table 2.1: Showing tabulated values of u_i, $\eta_i = \pi u_i$, $\mathrm{Sin}(\eta_i)$ from which we get $I = \frac{1}{N}\sum_{i=1}^{N}\frac{F(x_i)}{p(x_i)} = \frac{1}{N}\sum_{i=1}^{N}\frac{F(\eta_i)}{1/\pi} = \frac{\pi}{N}\sum_{i=1}^{N}\mathrm{Sin}(\eta_i)$ as an approximate value of the integral $I = \int_a^b F(x)dx = \int_0^\pi \mathrm{Sin}(x)dx = 1.973$ whereas the exact value is 2. Using $p(x) = 1/\pi$.

i	u_i	$\eta_i = \pi u_i$	$\mathrm{Sin}(\eta_i)$	i	u_i	$\eta_i = \pi u_i$	$\mathrm{Sin}(\eta_i)$
1	0.169	0.531	0.506	26	0.225	0.707	0.649
2	0.050	0.157	0.156	27	0.687	2.158	0.832
3	0.844	2.652	0.471	28	0.060	0.188	0.187
4	0.182	0.572	0.541	29	0.070	0.220	0.218
5	0.599	1.882	0.952	30	0.835	2.623	0.495
6	0.985	3.094	0.047	31	0.466	1.464	0.994
7	0.726	2.281	0.758	32	0.271	0.851	0.752
8	0.148	0.465	0.448	33	0.813	2.554	0.554
9	0.788	2.476	0.618	34	0.376	1.181	0.925
10	0.698	2.193	0.813	35	0.983	3.088	0.053
11	0.010	0.031	0.031	36	0.181	0.569	0.538
12	0.397	1.247	0.948	37	0.583	1.832	0.966
13	0.852	2.677	0.448	38	0.612	1.923	0.939
14	0.761	2.391	0.682	39	0.468	1.470	0.995
15	0.309	0.971	0.825	40	0.614	1.929	0.937
16	0.113	0.355	0.348	41	0.768	2.413	0.666
17	0.212	0.666	0.618	42	0.811	2.548	0.559
18	0.809	2.542	0.565	43	0.461	1.448	0.993
19	0.084	0.264	0.261	44	0.985	3.094	0.047
20	0.276	0.867	0.762	45	0.735	2.309	0.740
21	0.243	0.763	0.691	46	0.559	1.756	0.983
22	0.989	3.107	0.035	47	0.884	2.777	0.356
23	0.508	1.596	1.000	48	0.229	0.719	0.659
24	0.817	2.567	0.544	49	0.920	2.890	0.249
25	0.174	0.547	0.520	50	0.524	1.646	0.997

i	u_i	$\eta_i = \pi\, u_i$	$\mathrm{Sin}(\eta_i)$	i	u_i	$\eta_i = \pi\, u_i$	$\mathrm{Sin}(\eta_i)$
51	0.393	1.235	0.944	76	0.231	0.726	0.664
52	0.306	0.961	0.820	77	0.836	2.626	0.493
53	0.819	2.573	0.538	78	0.167	0.525	0.501
54	0.620	1.948	0.930	79	0.545	1.712	0.990
55	0.875	2.749	0.383	80	0.748	2.350	0.712
56	0.773	2.428	0.654	81	0.440	1.382	0.982
57	0.492	1.546	1.000	82	0.674	2.117	0.854
58	0.332	1.043	0.864	83	0.287	0.902	0.784
59	0.185	0.581	0.549	84	0.052	0.163	0.163
60	0.565	1.775	0.979	85	0.940	2.953	0.187
61	0.434	1.363	0.979	86	0.679	2.133	0.846
62	0.665	2.089	0.869	87	0.910	2.859	0.279
63	0.516	1.621	0.999	88	0.254	0.798	0.716
64	0.659	2.070	0.878	89	0.236	0.741	0.675
65	0.709	2.227	0.792	90	0.854	2.683	0.443
66	0.515	1.618	0.999	91	0.877	2.755	0.377
67	0.683	2.146	0.839	92	0.483	1.517	0.999
68	0.092	0.289	0.285	93	0.117	0.368	0.359
69	0.417	1.310	0.966	94	0.847	2.661	0.462
70	0.598	1.879	0.953	95	0.201	0.631	0.590
71	0.682	2.143	0.841	96	0.294	0.924	0.798
72	0.948	2.978	0.163	97	0.071	0.223	0.221
73	0.125	0.393	0.383	98	0.065	0.204	0.203
74	0.149	0.468	0.451	99	0.740	2.325	0.729
75	0.246	0.773	0.698	100	0.954	2.997	0.144
						Sum = 62.800	
						$I = 1.973$	

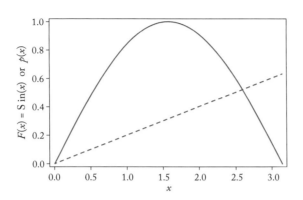

Figure 2.2: Showing $F(x) = \text{Sin}(x)$ as undashed curve and $p(x) = 2x/\pi^2$ as dashed curve obtained using the command
```
Plot[{Sin[x],(2*x)/(Pi^2)},{x,0,Pi},Frame->True,FrameLabel->
{"x","F(x) = Sin(x) or p(x)"},PlotStyle->
{{Black},{Dashed,Black}}]
```
in Mathematica 6.0.

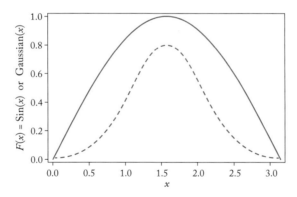

Figure 2.3: Showing $F(x) = \text{Sin}(x)$ as undashed curve and Gaussian $p(x) = \frac{1}{\sigma\sqrt{2\pi}}$ $\text{Exp}\left[-\frac{1}{2}\left(\frac{x-a}{\sigma}\right)^2\right]$ as dashed curve obtained using the command
```
sig=0.5;a=Pi/2;
p1=Plot[Sin[x],{x,0,Pi},PlotStyle->{Black},Frame->True,
FrameLabel->{"x","F(x) = Sin(x) or Gaussian (x)"}];
p2=Plot[(1/(sig*Sqrt[2*Pi]))*Exp[-0.5*((x-a)/sig)^2],{x,0,Pi},
PlotStyle->{Dashed,Black}];Show[p1,p2]
```
in Mathematica 6.0.

Table 2.2: Showing tabulated values of u_i, $\eta_i = \pi u_i^{1/2}$, $\text{Sin}(\eta_i)/\eta_i$ from which we get $I = \frac{1}{N}\sum_{i=1}^{N} \frac{F(x_i)}{p(x_i)} = \frac{1}{N}\sum_{i=1}^{N} \frac{\text{Sin}(\eta_i)}{2\eta_i/\pi^2} = \frac{\pi^2}{2N}\sum_{i=1}^{N} \frac{\text{Sin}(\eta_i)}{\eta_i}$ as an approximate value of the integral $I = \int_a^b F(x)dx = \int_0^\pi \text{Sin}(x)dx = 1.947$ whereas the exact value is 2. Using $p(x) = 2x/\pi^2$.

i	u_i	$\eta_i = \pi u_i^{1/2}$	$\text{Sin}(\eta_i)/\eta_i$	i	u_i	$\eta_i = \pi u_i^{1/2}$	$\text{Sin}(\eta_i)/\eta_i$
1	0.169	1.291	0.744	26	0.225	1.490	0.669
2	0.050	0.702	0.920	27	0.687	2.604	0.197
3	0.844	2.886	0.088	28	0.060	0.770	0.904
4	0.182	1.340	0.726	29	0.070	0.831	0.889
5	0.599	2.431	0.268	30	0.835	2.871	0.093
6	0.985	3.118	0.008	31	0.466	2.145	0.392
7	0.726	2.677	0.167	32	0.271	1.635	0.610
8	0.148	1.209	0.774	33	0.813	2.833	0.107
9	0.788	2.789	0.124	34	0.376	1.926	0.487
10	0.698	2.625	0.188	35	0.983	3.115	0.009
11	0.010	0.314	0.984	36	0.181	1.337	0.728
12	0.397	1.979	0.464	37	0.583	2.399	0.282
13	0.852	2.900	0.083	38	0.612	2.458	0.257
14	0.761	2.741	0.142	39	0.468	2.149	0.390
15	0.309	1.746	0.564	40	0.614	2.462	0.255
16	0.113	1.056	0.824	41	0.768	2.753	0.138
17	0.212	1.447	0.686	42	0.811	2.829	0.109
18	0.809	2.826	0.110	43	0.461	2.133	0.397
19	0.084	0.911	0.867	44	0.985	3.118	0.008
20	0.276	1.650	0.604	45	0.735	2.693	0.161
21	0.243	1.549	0.646	46	0.559	2.349	0.303
22	0.989	3.124	0.006	47	0.884	2.954	0.063
23	0.508	2.239	0.351	48	0.229	1.503	0.664
24	0.817	2.840	0.105	49	0.920	3.013	0.042
25	0.174	1.310	0.737	50	0.524	2.274	0.335

i	u_i	$\eta_i = \pi\, u_i^{1/2}$	$\mathrm{Sin}(\eta_i)/\eta_i$	i	u_i	$\eta_i = \pi\, u_i^{1/2}$	$\mathrm{Sin}(\eta_i)/\eta_i$
51	0.393	1.969	0.468	76	0.231	1.510	0.661
52	0.306	1.738	0.567	77	0.836	2.872	0.093
53	0.819	2.843	0.103	78	0.167	1.284	0.747
54	0.620	2.474	0.250	79	0.545	2.319	0.316
55	0.875	2.939	0.069	80	0.748	2.717	0.152
56	0.773	2.762	0.134	81	0.440	2.084	0.418
57	0.492	2.204	0.366	82	0.674	2.579	0.207
58	0.332	1.810	0.537	83	0.287	1.683	0.590
59	0.185	1.351	0.722	84	0.052	0.716	0.917
60	0.565	2.361	0.298	85	0.940	3.046	0.031
61	0.434	2.070	0.424	86	0.679	2.589	0.203
62	0.665	2.562	0.214	87	0.910	2.997	0.048
63	0.516	2.257	0.343	88	0.254	1.583	0.632
64	0.659	2.550	0.219	89	0.236	1.526	0.655
65	0.709	2.645	0.180	90	0.854	2.903	0.081
66	0.515	2.255	0.344	91	0.877	2.942	0.067
67	0.683	2.596	0.200	92	0.483	2.183	0.375
68	0.092	0.953	0.855	93	0.117	1.075	0.818
69	0.417	2.029	0.442	94	0.847	2.891	0.086
70	0.598	2.429	0.269	95	0.201	1.408	0.701
71	0.682	2.594	0.201	96	0.294	1.703	0.582
72	0.948	3.059	0.027	97	0.071	0.837	0.887
73	0.125	1.111	0.807	98	0.065	0.801	0.896
74	0.149	1.213	0.772	99	0.740	2.703	0.157
75	0.246	1.558	0.642	100	0.954	3.068	0.024
						Sum = 39.461	
						$I = 1.947$	

$p(x)$. As Figure 2.3 shows, the Gaussian probability density function $p(x)$ with $a = \pi/2$ and $\sigma = 0.5$ is not too different from the function $\text{Sin}(x)$. Hence, it is ideal for use as $p(x)$.

We now generate random variable G obeying

$$p(G) = \frac{1}{\sigma\sqrt{2\pi}} \text{Exp}\left[-\frac{1}{2}\left(\frac{G-a}{\sigma}\right)^2\right]$$

with $a = \pi/2$ and $\sigma = 0.5$ using the tabulated values of random variable g of Table 1.4 obeying

$$p(g) = \frac{1}{\sigma\sqrt{2\pi}} \text{Exp}\left[-\frac{1}{2}\left(\frac{g-a}{\sigma}\right)^2\right]$$

for $a = 0$ and $\sigma = 1$. The connection is $G = a + \sigma g = \pi/2 + 0.5g$.

Using Table 2.3 in Microsoft Excel, we have evaluated the integral with the result 1.950 rather than 2. The difference is attributable to the fact that $F(x)$ and $p(x)$ used are not proportional to each other.

2.2 EVALUATION OF DEFINITE INTEGRALS USING THE MONTE CARLO METHOD: EXAMPLE II

As Example II, we now take up the integral

$$I = \int_a^b F(x)dx = \int_{-\pi/2}^{+\pi/2} \text{Cos}(x)dx, \tag{2.4}$$

where $F(x) = \text{Cos}(x)$. We re-write Equation (2.4) as

$$I = \int_a^b \frac{F(x)}{p(x)} p(x)dx \tag{2.5}$$

which, as discussed in Section 1.7, implies that average value of F/p is the value of the integral, i.e.,

$$I = \frac{1}{N} \sum_{i=1}^{N} \frac{F(x_i)}{p(x_i)}, \tag{2.6}$$

where x_i's are random values of x in the interval $a < x < b$ obeying probability density function $p(x)$.

Here, $p(x)$ is a suitable probability density function which, as demonstrated in Section 1.7, at least should follow $F(x)$, if not be proportional to $F(x)$. See Figure 2.4 in which we have plotted $F(x) = \text{Cos}(x)$ along with normalized probability density function of a uniform random variable in the interval $-\pi/2$ to $+\pi/2$ given by $p(x) = 1/\pi$. Let $p(x) = C$, a constant. Normalization requires that $\int_{-\pi/2}^{+\pi/2} C dx = 1$ which gives $C = 1/\pi$. Uniform $p(x) = 1/\pi$ is not too different from a slowly varying function of x like $\text{Cos}(x)$. Hence, $p(x) = 1/\pi$ is a good choice.

Table 2.3: Showing tabulated values of u_i, g_i, G_i, $Sin(G_i)/p_i$ from which we get $I = \frac{1}{N} \sum_{i=1}^{N} \frac{F(x_i)}{p(x_i)} = \frac{1}{N} \sum_{i=1}^{N} \frac{Sin(G_i)}{p_i} = \frac{1}{N} \sum_{i=1}^{N} \frac{Sin(G_i)}{\frac{1}{\sigma\sqrt{2\pi}} Exp\left[-\frac{1}{2}\left(\frac{G_i-a}{\sigma}\right)^2\right]}$ as an approximate value of the integral $I = \int_a^b F(x)dx = \int_0^\pi Sin(x)dx = 1.950$ whereas the exact value is 2. Using $p(x) = \frac{1}{\sigma\sqrt{2\pi}} Exp\left[-\frac{1}{2}\left(\frac{x-a}{\sigma}\right)^2\right]$.

i	u_i	g_i	G_i	$SinG_i$	p_i	$(SinG_i)/p_i$
1	0.169	−0.955	1.093	0.888	0.506	1.756
2	0.050	−1.635	0.753	0.684	0.210	3.263
3	0.844	1.015	2.078	0.874	0.477	1.833
4	0.182	−0.905	1.118	0.899	0.530	1.698
5	0.599	0.255	1.698	0.992	0.772	1.284
6	0.985	2.195	2.668	0.456	0.072	6.354
7	0.726	0.605	1.873	0.955	0.664	1.437
8	0.148	−1.040	1.051	0.868	0.465	1.868
9	0.788	0.805	1.973	0.920	0.577	1.594
10	0.698	0.520	1.831	0.966	0.697	1.387
11	0.010	−2.295	0.423	0.411	0.057	7.168
12	0.397	−0.255	1.443	0.992	0.772	1.284
13	0.852	1.050	2.096	0.865	0.460	1.882
14	0.761	0.715	1.928	0.937	0.618	1.516
15	0.309	−0.495	1.323	0.970	0.706	1.373
16	0.113	−1.205	0.968	0.824	0.386	2.134
17	0.212	−0.795	1.173	0.922	0.582	1.585
18	0.809	0.880	2.011	0.905	0.542	1.670
19	0.084	−1.370	0.886	0.774	0.312	2.481
20	0.276	−0.590	1.276	0.957	0.670	1.427
21	0.243	−0.695	1.223	0.940	0.627	1.500
22	0.989	2.325	2.733	0.397	0.053	7.425
23	0.508	0.020	1.581	1.000	0.798	1.254
24	0.817	0.910	2.026	0.898	0.527	1.703
25	0.174	−0.935	1.103	0.893	0.515	1.732

i	u_i	g_i	G_i	$\mathrm{Sin}\,G_i$	p_i	$(\mathrm{Sin}\,G_i)/p_i$
26	0.225	−0.750	1.196	0.931	0.602	1.545
27	0.687	0.490	1.816	0.970	0.708	1.371
28	0.060	−1.540	0.801	0.718	0.244	2.945
29	0.070	−1.465	0.838	0.744	0.273	2.725
30	0.835	0.980	2.061	0.882	0.494	1.787
31	0.466	−0.085	1.528	0.999	0.795	1.257
32	0.271	−0.605	1.268	0.955	0.664	1.437
33	0.813	0.895	2.018	0.902	0.535	1.686
34	0.376	−0.315	1.413	0.988	0.759	1.301
35	0.983	2.145	2.643	0.478	0.080	5.978
36	0.181	−0.905	1.118	0.899	0.530	1.698
37	0.583	0.210	1.676	0.994	0.780	1.274
38	0.612	0.285	1.713	0.990	0.766	1.292
39	0.468	−0.080	1.531	0.999	0.795	1.256
40	0.614	0.290	1.716	0.990	0.765	1.293
41	0.768	0.735	1.938	0.933	0.609	1.532
42	0.811	0.885	2.013	0.904	0.539	1.676
43	0.461	−0.095	1.523	0.999	0.794	1.258
44	0.985	2.195	2.668	0.456	0.072	6.354
45	0.735	0.630	1.886	0.951	0.654	1.453
46	0.559	0.150	1.646	0.997	0.789	1.264
47	0.884	1.200	2.171	0.825	0.388	2.125
48	0.229	−0.740	1.201	0.932	0.607	1.537
49	0.920	1.415	2.278	0.760	0.293	2.592
50	0.524	0.060	1.601	1.000	0.796	1.255

i	u_i	g_i	G_i	$\mathrm{Sin}G_i$	p_i	$(\mathrm{Sin}G_i)/p_i$
51	0.393	−0.270	1.436	0.991	0.769	1.288
52	0.306	−0.505	1.318	0.968	0.702	1.379
53	0.819	0.915	2.028	0.897	0.525	1.709
54	0.620	0.310	1.726	0.988	0.760	1.299
55	0.875	1.155	2.148	0.838	0.410	2.046
56	0.773	0.750	1.946	0.931	0.602	1.545
57	0.492	−0.020	1.561	1.000	0.798	1.254
58	0.332	−0.430	1.356	0.977	0.727	1.343
59	0.185	−0.890	1.126	0.903	0.537	1.681
60	0.565	0.165	1.653	0.997	0.787	1.266
61	0.434	−0.165	1.488	0.997	0.787	1.266
62	0.665	0.430	1.786	0.977	0.727	1.343
63	0.516	0.040	1.591	1.000	0.797	1.254
64	0.659	0.410	1.776	0.979	0.734	1.335
65	0.709	0.555	1.848	0.962	0.684	1.406
66	0.515	0.040	1.591	1.000	0.797	1.254
67	0.683	0.480	1.811	0.971	0.711	1.366
68	0.092	−1.320	0.911	0.790	0.334	2.366
69	0.417	−0.205	1.468	0.995	0.781	1.273
70	0.598	0.250	1.696	0.992	0.773	1.283
71	0.682	0.475	1.808	0.972	0.713	1.364
72	0.948	1.635	2.388	0.684	0.210	3.263
73	0.125	−1.145	0.998	0.841	0.414	2.029
74	0.149	−1.035	1.053	0.869	0.467	1.861
75	0.246	−0.685	1.228	0.942	0.631	1.493

i	u_i	g_i	G_i	$\mathrm{Sin}\,G_i$	p_i	$(\mathrm{Sin}\,G_i)/p_i$
76	0.231	−0.730	1.206	0.934	0.611	1.528
77	0.836	0.985	2.063	0.881	0.491	1.794
78	0.167	−0.960	1.091	0.887	0.503	1.762
79	0.545	0.115	1.628	0.998	0.793	1.260
80	0.748	0.670	1.906	0.944	0.637	1.481
81	0.440	−0.150	1.496	0.997	0.789	1.264
82	0.674	0.455	1.798	0.974	0.719	1.354
83	0.287	−0.560	1.291	0.961	0.682	1.409
84	0.052	−1.615	0.763	0.691	0.217	3.192
85	0.940	1.565	2.353	0.709	0.234	3.024
86	0.679	0.470	1.806	0.973	0.714	1.361
87	0.910	1.350	2.246	0.781	0.321	2.434
88	0.254	−0.660	1.241	0.946	0.642	1.474
89	0.236	−0.715	1.213	0.937	0.618	1.516
90	0.854	1.060	2.101	0.863	0.455	1.897
91	0.877	1.165	2.153	0.835	0.405	2.063
92	0.483	−0.040	1.551	1.000	0.797	1.254
93	0.117	−1.185	0.978	0.830	0.395	2.098
94	0.847	1.030	2.086	0.870	0.469	1.854
95	0.201	−0.835	1.153	0.914	0.563	1.624
96	0.294	−0.540	1.301	0.964	0.690	1.398
97	0.071	−1.460	0.841	0.745	0.275	2.711
98	0.065	−1.505	0.818	0.730	0.257	2.839
99	0.740	0.645	1.893	0.948	0.648	1.464
100	0.954	1.695	2.418	0.662	0.190	3.489
						Sum = 195.012
						I = 1.950

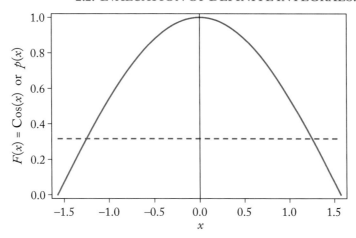

Figure 2.4: Showing $F(x) = \text{Cos}(x)$ as undashed curve and $p(x) = 1/\pi$ as dashed curve obtained using the command
```
Plot[{Cos[x],1/Pi},{x,-Pi/2,Pi/2},
Frame->True,FrameLabel->{"x","F(x) = Cos(x) or p(x)"},
PlotStyle->{{Black},{Dashed,Black}}]
```
in Mathematica 6.0.

We now generate random variable η obeying $p(x) = 1/\pi$ in the interval $-\pi/2 < x < +\pi/2$ using Equation (1.4): $\eta = a + (b - a)u$ where u is random variable (of Table 1.1) of uniform probability density function in the interval $0 < x < 1$. Thus, we have $\eta = \pi u - \pi/2$. That $\eta = \pi u - \pi/2$ can be calculated using $\int_a^\eta p(x)dx = u$ which gives

$$\int_{-\pi/2}^{\eta} (1/\pi)dx = u \quad \text{or,} \quad \eta = \pi u - \pi/2.$$

Using Table 2.4 in Microsoft Excel, we have evaluated the integral with the result 1.973 rather than 2. The difference is attributable to the fact that $F(x)$ and $p(x)$ used are not proportional to each other.

We now take up a linear variation for $p(x)$ given by $p(x) = Cx + d$ where C and d are parameters such that slope $C = $ intercept d on y axis $/(\pi/2)$; hence, $d = C\pi/2$. As such, $p(x) = C(x + \pi/2)$. Normalization of $p(x)$ requires $\int_{-\pi/2}^{+\pi/2} p(x)dx = 1$ or $C = 2/\pi^2$. As such, normalized probability density function is $p(x) = 2x/\pi^2 + 1/\pi$. See Figure 2.5 in which we have plotted $F(x) = \text{Cos}(x)$ along with the normalized linear probability density function $p(x) = 2x/\pi^2 + 1/\pi$. The linear probability density function $p(x) = 2x/\pi^2 + 1/\pi$ is not too different from a slowly varying function of x like $\text{Cos}(x)$. Hence, it is a good choice.

Table 2.4: Showing tabulated values of u_i, $\eta_i = \pi u_i - \pi/2$, $\mathrm{Cos}(\eta_i)$ from which we get $I = \frac{1}{N}\sum_{i=1}^{N}\frac{F(x_i)}{p(x_i)} = \frac{1}{N}\sum_{i=1}^{N}\frac{F(\eta_i)}{1/\pi} = \frac{\pi}{N}\sum_{i=1}^{N}\mathrm{Cos}(\eta_i)$ as an approximate value of the integral $I = \int_a^b F(x)dx = \int_{-\pi/2}^{+\pi/2}\mathrm{Cos}(x)dx = 1.973$ whereas the exact value is 2. Using $p(x) = 1/\pi$.

i	u_i	$\eta_i = \pi u_i - \pi/2$	$\mathrm{Cos}(\eta_i)$	i	u_i	$\eta_i = \pi u_i - \pi/2$	$\mathrm{Cos}(\eta_i)$
1	0.169	−1.040	0.506	26	0.225	−0.864	0.649
2	0.050	−1.414	0.156	27	0.687	0.587	0.832
3	0.844	1.081	0.471	28	0.060	−1.382	0.187
4	0.182	−0.999	0.541	29	0.070	−1.351	0.218
5	0.599	0.311	0.952	30	0.835	1.052	0.495
6	0.985	1.524	0.047	31	0.466	−0.107	0.994
7	0.726	0.710	0.758	32	0.271	−0.719	0.752
8	0.148	−1.106	0.448	33	0.813	0.983	0.554
9	0.788	0.905	0.618	34	0.376	−0.390	0.925
10	0.698	0.622	0.813	35	0.983	1.517	0.053
11	0.010	−1.539	0.031	36	0.181	−1.002	0.538
12	0.397	−0.324	0.948	37	0.583	0.261	0.966
13	0.852	1.106	0.448	38	0.612	0.352	0.939
14	0.761	0.820	0.682	39	0.468	−0.101	0.995
15	0.309	−0.600	0.825	40	0.614	0.358	0.937
16	0.113	−1.216	0.348	41	0.768	0.842	0.666
17	0.212	−0.905	0.618	42	0.811	0.977	0.559
18	0.809	0.971	0.565	43	0.461	−0.123	0.993
19	0.084	−1.307	0.261	44	0.985	1.524	0.047
20	0.276	−0.704	0.762	45	0.735	0.738	0.740
21	0.243	−0.807	0.691	46	0.559	0.185	0.983
22	0.989	1.536	0.035	47	0.884	1.206	0.356
23	0.508	0.025	1.000	48	0.229	−0.851	0.659
24	0.817	0.996	0.544	49	0.920	1.319	0.249
25	0.174	−1.024	0.520	50	0.524	0.075	0.997

i	u_i	$\eta_i = \pi u_i - \pi/2$	$\cos(\eta_i)$	i	u_i	$\eta_i = \pi u_i - \pi/2$	$\cos(\eta_i)$
51	0.393	−0.336	0.944	76	0.231	−0.845	0.664
52	0.306	−0.609	0.820	77	0.836	1.056	0.493
53	0.819	1.002	0.538	78	0.167	−1.046	0.501
54	0.620	0.377	0.930	79	0.545	0.141	0.990
55	0.875	1.178	0.383	80	0.748	0.779	0.712
56	0.773	0.858	0.654	81	0.440	−0.188	0.982
57	0.492	−0.025	1.000	82	0.674	0.547	0.854
58	0.332	−0.528	0.864	83	0.287	−0.669	0.784
59	0.185	−0.990	0.549	84	0.052	−1.407	0.163
60	0.565	0.204	0.979	85	0.940	1.382	0.187
61	0.434	−0.207	0.979	86	0.679	0.562	0.846
62	0.665	0.518	0.869	87	0.910	1.288	0.279
63	0.516	0.050	0.999	88	0.254	−0.773	0.716
64	0.659	0.500	0.878	89	0.236	−0.829	0.675
65	0.709	0.657	0.792	90	0.854	1.112	0.443
66	0.515	0.047	0.999	91	0.877	1.184	0.377
67	0.683	0.575	0.839	92	0.483	−0.053	0.999
68	0.092	−1.282	0.285	93	0.117	−1.203	0.359
69	0.417	−0.261	0.966	94	0.847	1.090	0.462
70	0.598	0.308	0.953	95	0.201	−0.939	0.590
71	0.682	0.572	0.841	96	0.294	−0.647	0.798
72	0.948	1.407	0.163	97	0.071	−1.348	0.221
73	0.125	−1.178	0.383	98	0.065	−1.367	0.203
74	0.149	−1.103	0.451	99	0.740	0.754	0.729
75	0.246	−0.798	0.698	100	0.954	1.426	0.144

Sum = 62.800

I = 1.973

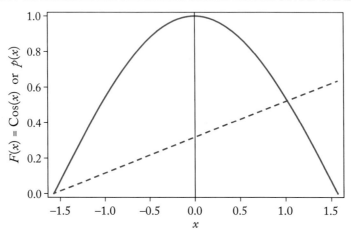

Figure 2.5: Showing $F(x) = \text{Cos}(x)$ as undashed curve and $p(x) = 2x/\pi^2 + 1/\pi$ as dashed curve obtained using the command

```
Plot[{Cos[x],2*x/Pi^2+1/Pi},{x,-Pi/2,Pi/2},Frame->True,
FrameLabel->{"x","F(x) = Cos(x) or p(x)"},
PlotStyle->{{Black},{Dashed,Black}}]
```

in Mathematica 6.0.

We now generate random variable η obeying $p(x) = 2x/\pi^2 + 1/\pi$ in the interval $-\pi/2 < x < +\pi/2$ using Equation (1.2):

$$\int_a^\eta p(x)dx = u \quad \text{or,} \quad \int_{-\pi/2}^\eta p(x)dx = u,$$

where u is random variable (of Table 1.1) of uniform probability density function in the interval $0 < x < 1$. Thus, we have η given by roots of the quadratic equation

$$\eta^2 + \pi\eta + (\pi^2/4 - \pi^2 u) = 0$$

which are

$$\eta = -\pi/2 \pm \pi u^{1/2}.$$

Using $\eta = -\pi/2 + \pi u^{1/2}$ in Microsoft Excel we have Table 2.5 in which we have evaluated the integral with the result 1.947 rather than 2. The difference is attributable to the fact that $F(x)$ and $p(x)$ used are not proportional to each other. Table 2.6 uses $\eta = -\pi/2 - \pi u^{1/2}$.

We now take up a normalized *Gaussian* variation for $p(x)$ given by $p(x) = \frac{1}{\sigma\sqrt{2\pi}}\text{Exp}\left[-\frac{1}{2}\left(\frac{x-a}{\sigma}\right)^2\right]$ where a is average and σ is variance of x. See Figure 2.6 in which we have plotted $F(x) = \text{Cos}(x)$ along with the normalized Gaussian probability density function $p(x)$.

Table 2.5: Showing tabulated values of u_i, $\eta_i = -\pi/2 + \pi u_i^{1/2}$, $\mathrm{Cos}(\eta_i)/p_i$ from which we get $I = \frac{1}{N} \sum_{i=1}^{N} \frac{F(x_i)}{p(x_i)} = \frac{1}{N} \sum_{i=1}^{N} \frac{\mathrm{Cos}(\eta_i)}{p_i} = \frac{1}{N} \sum_{i=1}^{N} \frac{\mathrm{Cos}(\eta_i)}{1/\pi + 2\eta_i/\pi^2}$ as an approximate value of the integral $I = \int_a^b F(x)dx = \int_{-\pi/2}^{+\pi/2} \mathrm{Cos}(x)dx = 1.947$ whereas the exact value is 2. Using $p(x) = 2x/\pi^2 + 1/\pi$.

i	u_i	$\eta_i = -\pi/2 + \pi u_i^{1/2}$	$\mathrm{Cos}(\eta_i)/p_i = \mathrm{Cos}(\eta_i)/(1/\pi + 2\eta_i/\pi^2)$
1	0.169	−0.279	3.673
2	0.050	−0.868	4.539
3	0.844	1.315	0.432
4	0.182	−0.231	3.585
5	0.599	0.861	1.323
6	0.985	1.547	0.037
7	0.726	1.106	0.826
8	0.148	−0.362	3.818
9	0.788	1.218	0.611
10	0.698	1.054	0.929
11	0.010	−1.257	4.854
12	0.397	0.409	2.288
13	0.852	1.329	0.407
14	0.761	1.170	0.703
15	0.309	0.176	2.782
16	0.113	−0.515	4.067
17	0.212	−0.124	3.385
18	0.809	1.255	0.543
19	0.084	−0.660	4.281
20	0.276	0.080	2.980
21	0.243	−0.022	3.186
22	0.989	1.553	0.027
23	0.508	0.668	1.730
24	0.817	1.269	0.517
25	0.174	−0.260	3.639

i	u_i	$\eta_i = -\pi/2 + \pi u_i^{1/2}$	$\mathrm{Cos}(\eta_i)/p_i = \mathrm{Cos}(\eta_i)/(1/\pi + 2\eta_i/\pi^2)$
26	0.225	−0.081	3.301
27	0.687	1.033	0.971
28	0.060	−0.801	4.462
29	0.070	−0.740	4.386
30	0.835	1.300	0.460
31	0.466	0.574	1.933
32	0.271	0.065	3.011
33	0.813	1.262	0.530
34	0.376	0.356	2.401
35	0.983	1.544	0.042
36	0.181	−0.234	3.591
37	0.583	0.828	1.391
38	0.612	0.887	1.269
39	0.468	0.578	1.923
40	0.614	0.891	1.260
41	0.768	1.182	0.679
42	0.811	1.258	0.536
43	0.461	0.562	1.957
44	0.985	1.547	0.037
45	0.735	1.123	0.794
46	0.559	0.778	1.496
47	0.884	1.383	0.312
48	0.229	−0.067	3.275
49	0.920	1.443	0.210
50	0.524	0.703	1.655

i	u_i	$\eta_i = -\pi/2 + \pi u_i^{1/2}$	$\mathrm{Cos}(\eta_i)/p_i = \mathrm{Cos}(\eta_i)/(1/\pi + 2\eta_i/\pi^2)$
51	0.393	0.399	2.309
52	0.306	0.167	2.800
53	0.819	1.272	0.510
54	0.620	0.903	1.236
55	0.875	1.368	0.338
56	0.773	1.191	0.662
57	0.492	0.633	1.806
58	0.332	0.239	2.648
59	0.185	−0.220	3.564
60	0.565	0.791	1.470
61	0.434	0.499	2.094
62	0.665	0.991	1.055
63	0.516	0.686	1.692
64	0.659	0.980	1.079
65	0.709	1.074	0.888
66	0.515	0.684	1.697
67	0.683	1.026	0.986
68	0.092	−0.618	4.221
69	0.417	0.458	2.182
70	0.598	0.859	1.327
71	0.682	1.024	0.990
72	0.948	1.488	0.133
73	0.125	−0.460	3.981
74	0.149	−0.358	3.811
75	0.246	−0.013	3.167

i	u_i	$\eta_i = -\pi/2 + \pi u_i^{1/2}$	$\text{Cos}(\eta_i)/p_i = \text{Cos}(\eta_i)/(1/\pi + 2\eta_i/\pi^2)$
76	0.231	−0.061	3.262
77	0.836	1.302	0.457
78	0.167	−0.287	3.687
79	0.545	0.748	1.559
80	0.748	1.146	0.748
81	0.440	0.513	2.063
82	0.674	1.008	1.020
83	0.287	0.112	2.914
84	0.052	−0.854	4.523
85	0.940	1.475	0.155
86	0.679	1.018	1.001
87	0.910	1.426	0.237
88	0.254	0.013	3.117
89	0.236	−0.045	3.230
90	0.854	1.332	0.401
91	0.877	1.371	0.332
92	0.483	0.613	1.849
93	0.117	−0.496	4.038
94	0.847	1.320	0.423
95	0.201	−0.162	3.458
96	0.294	0.133	2.872
97	0.071	−0.734	4.378
98	0.065	−0.770	4.424
99	0.740	1.132	0.776
100	0.954	1.498	0.117
			Sum = 194.736
			$I = 1.947$

Table 2.6: Showing tabulated values of u_i, $\eta_i = -\pi/2 - \pi u_i^{1/2}$, $\text{Cos}(\eta_i)/p_i$ from which we get $I = \frac{1}{N} \sum_{i=1}^{N} \frac{F(x_i)}{p(x_i)} = \frac{1}{N} \sum_{i=1}^{N} \frac{\text{Cos}(\eta_i)}{p_i} = \frac{1}{N} \sum_{i=1}^{N} \frac{\text{Cos}(\eta_i)}{1/\pi + 2\eta_i/\pi^2}$ as an approximate value of the integral $I = \int_a^b F(x)dx = \int_{-\pi/2}^{+\pi/2} \text{Cos}(x)dx = 1.947$ whereas the exact value is 2. Using $p(x) = 1/\pi + 2x/\pi^2$.

i	u_i	$\eta_i = -\pi/2 - \pi u_i^{1/2}$	$\text{Cos}(\eta_i)/p_i = \text{Cos}(\eta_i)/(1/\pi + 2\eta_i/\pi^2)$
1	0.169	−2.862	3.673
2	0.050	−2.273	4.539
3	0.844	−4.457	0.432
4	0.182	−2.911	3.585
5	0.599	−4.002	1.323
6	0.985	−4.689	0.037
7	0.726	−4.248	0.826
8	0.148	−2.779	3.818
9	0.788	−4.360	0.611
10	0.698	−4.195	0.929
11	0.010	−1.885	4.854
12	0.397	−3.550	2.288
13	0.852	−4.471	0.407
14	0.761	−4.311	0.703
15	0.309	−3.317	2.782
16	0.113	−2.627	4.067
17	0.212	−3.017	3.385
18	0.809	−4.396	0.543
19	0.084	−2.481	4.281
20	0.276	−3.221	2.980
21	0.243	−3.119	3.186
22	0.989	−4.695	0.027
23	0.508	−3.810	1.730
24	0.817	−4.410	0.517
25	0.174	−2.881	3.639

i	u_i	$\eta_i = -\pi/2 - \pi u_i^{1/2}$	$\text{Cos}(\eta_i)/p_i = \text{Cos}(\eta_i)/(1/\pi + 2\eta_i/\pi^2)$
26	0.225	−3.061	3.301
27	0.687	−4.175	0.971
28	0.060	−2.340	4.462
29	0.070	−2.402	4.386
30	0.835	−4.442	0.460
31	0.466	−3.715	1.933
32	0.271	−3.206	3.011
33	0.813	−4.403	0.530
34	0.376	−3.497	2.401
35	0.983	−4.686	0.042
36	0.181	−2.907	3.591
37	0.583	−3.970	1.391
38	0.612	−4.028	1.269
39	0.468	−3.720	1.923
40	0.614	−4.032	1.260
41	0.768	−4.324	0.679
42	0.811	−4.400	0.536
43	0.461	−3.704	1.957
44	0.985	−4.689	0.037
45	0.735	−4.264	0.794
46	0.559	−3.920	1.496
47	0.884	−4.525	0.312
48	0.229	−3.074	3.275
49	0.920	−4.584	0.209
50	0.524	−3.845	1.655

i	u_i	$\eta_i = -\pi/2 - \pi u_i^{1/2}$	$\mathrm{Cos}(\eta_i)/p_i = \mathrm{Cos}(\eta_i)/(1/\pi + 2\eta_i/\pi^2)$
51	0.393	−3.540	2.309
52	0.306	−3.309	2.800
53	0.819	−4.414	0.510
54	0.620	−4.044	1.236
55	0.875	−4.509	0.338
56	0.773	−4.333	0.662
57	0.492	−3.774	1.806
58	0.332	−3.381	2.648
59	0.185	−2.922	3.564
60	0.565	−3.932	1.470
61	0.434	−3.640	2.094
62	0.665	−4.133	1.055
63	0.516	−3.828	1.692
64	0.659	−4.121	1.079
65	0.709	−4.216	0.888
66	0.515	−3.825	1.697
67	0.683	−4.167	0.986
68	0.092	−2.524	4.221
69	0.417	−3.600	2.182
70	0.598	−4.000	1.327
71	0.682	−4.165	0.990
72	0.948	−4.630	0.133
73	0.125	−2.682	3.981
74	0.149	−2.783	3.811
75	0.246	−3.129	3.167

i	u_i	$\eta_i = -\pi/2 - \pi u_i^{1/2}$	$\mathrm{Cos}(\eta_i)/p_i = \mathrm{Cos}(\eta_i)/(1/\pi + 2\eta_i/\pi^2)$
76	0.231	−3.081	3.262
77	0.836	−4.443	0.457
78	0.167	−2.855	3.687
79	0.545	−3.890	1.559
80	0.748	−4.288	0.748
81	0.440	−3.655	2.063
82	0.674	−4.150	1.020
83	0.287	−3.254	2.914
84	0.052	−2.287	4.523
85	0.940	−4.617	0.155
86	0.679	−4.160	1.001
87	0.910	−4.568	0.237
88	0.254	−3.154	3.117
89	0.236	−3.097	3.230
90	0.854	−4.474	0.401
91	0.877	−4.513	0.332
92	0.483	−3.754	1.849
93	0.117	−2.645	4.038
94	0.847	−4.462	0.423
95	0.201	−2.979	3.458
96	0.294	−3.274	2.872
97	0.071	−2.408	4.378
98	0.065	−2.372	4.424
99	0.740	−4.273	0.776
100	0.954	−4.639	0.117
			Sum = 194.735
			I = 1.947

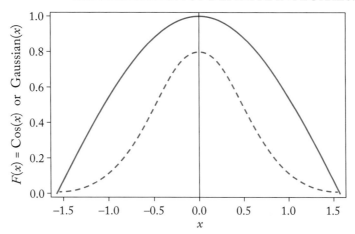

Figure 2.6: Showing $F(x) = \text{Cos}(x)$ as undashed curve and Gaussian $p(x) = \frac{1}{\sigma\sqrt{2\pi}}$ $\text{Exp}\left[-\frac{1}{2}\left(\frac{x-a}{\sigma}\right)^2\right]$ as dashed curve obtained using the command
sig=0.5; a=0;
p1=Plot[Cos[x],{x,-Pi/2,Pi/2},PlotStyle->{Black},Frame->True,
FrameLabel->{"x","F(x)= Cos (x) or Gaussian (x)"}];
p2=Plot[(1/(sig*Sqrt[2*Pi]))*Exp[-0.5*((x-a)/sig)^2],
{x,-Pi/2,Pi/2},PlotStyle->{Dashed,Black}]; Show[p1,p2]
in Mathematica 6.0.

As Figure 2.6 shows, the Gaussian probability density function $p(x)$ with $a = 0$ and $\sigma = 0.5$ is not too different from the function $\text{Cos}(x)$. Hence, it is ideal for use as $p(x)$.

We now generate random variable G obeying

$$p(G) = \frac{1}{\sigma\sqrt{2\pi}}\text{Exp}\left[-\frac{1}{2}\left(\frac{G-a}{\sigma}\right)^2\right]$$

with $a = 0$ and $\sigma = 0.5$ using the tabulated values of random variable g of Table 1.4 obeying

$$p(g) = \frac{1}{\sigma\sqrt{2\pi}}\text{Exp}\left[-\frac{1}{2}\left(\frac{g-a}{\sigma}\right)^2\right]$$

for $a = 0$ and $\sigma = 1$. The connection is $G = a + \sigma g = 0 + 0.5g$.

Using Table 2.7 in Microsoft Excel, we have evaluated the integral with the result 1.950 rather than 2. The difference is attributable to the fact that $F(x)$ and $p(x)$ used are not proportional to each other.

Table 2.7: Showing tabulated values of u_i, g_i, G_i, $\text{Cos}(G_i)/p_i$ from which we get $I = \frac{1}{N}\sum_{i=1}^{N}$ $\frac{F(x_i)}{p(x_i)} = \frac{1}{N}\sum_{i=1}^{N}\frac{\text{Cos}(G_i)}{p_i} = \frac{1}{N}\sum_{i=1}^{N}\frac{\text{Cos}(G_i)}{\frac{1}{\sigma\sqrt{2\pi}}\text{Exp}\left[-\frac{1}{2}\left(\frac{G_i-a}{\sigma}\right)^2\right]}$ as an approximate value of the integral $I = \int_a^b F(x)dx = \int_{-\pi/2}^{+\pi/2}\text{Cos}(x)dx = 1.950$ whereas the exact value is 2. Using $p(x) = \frac{1}{\sigma\sqrt{2\pi}}\text{Exp}\left[-\frac{1}{2}\left(\frac{x-a}{\sigma}\right)^2\right]$.

i	u_i	g_i	G_i	$\text{Cos}\,G_i$	p_i	$(\text{Cos}\,G_i)/p_i$
1	0.169	−0.955	−0.478	0.888	0.506	1.756
2	0.050	−1.635	−0.818	0.684	0.210	3.263
3	0.844	1.015	0.508	0.874	0.477	1.833
4	0.182	−0.905	−0.453	0.899	0.530	1.698
5	0.599	0.255	0.128	0.992	0.772	1.284
6	0.985	2.195	1.098	0.456	0.072	6.354
7	0.726	0.605	0.303	0.955	0.664	1.437
8	0.148	−1.040	−0.520	0.868	0.465	1.868
9	0.788	0.805	0.403	0.920	0.577	1.594
10	0.698	0.520	0.260	0.966	0.697	1.387
11	0.010	−2.295	−1.148	0.411	0.057	7.168
12	0.397	−0.255	−0.128	0.992	0.772	1.284
13	0.852	1.050	0.525	0.865	0.460	1.882
14	0.761	0.715	0.358	0.937	0.618	1.516
15	0.309	−0.495	−0.248	0.970	0.706	1.373
16	0.113	−1.205	−0.603	0.824	0.386	2.134
17	0.212	−0.795	−0.398	0.922	0.582	1.585
18	0.809	0.880	0.440	0.905	0.542	1.670
19	0.084	−1.370	−0.685	0.774	0.312	2.481
20	0.276	−0.590	−0.295	0.957	0.670	1.427
21	0.243	−0.695	−0.348	0.940	0.627	1.500
22	0.989	2.325	1.163	0.397	0.053	7.425
23	0.508	0.020	0.010	1.000	0.798	1.254
24	0.817	0.910	0.455	0.898	0.527	1.703
25	0.174	−0.935	−0.468	0.893	0.515	1.732

i	u_i	g_i	G_i	Cos G_i	p_i	(Cos G_i)/p_i
26	0.225	−0.750	−0.375	0.931	0.602	1.545
27	0.687	0.490	0.245	0.970	0.708	1.371
28	0.060	−1.540	−0.770	0.718	0.244	2.945
29	0.070	−1.465	−0.733	0.744	0.273	2.725
30	0.835	0.980	0.490	0.882	0.494	1.787
31	0.466	−0.085	−0.043	0.999	0.795	1.257
32	0.271	−0.605	−0.303	0.955	0.664	1.437
33	0.813	0.895	0.448	0.902	0.535	1.686
34	0.376	−0.315	−0.158	0.988	0.759	1.301
35	0.983	2.145	1.073	0.478	0.080	5.978
36	0.181	−0.905	−0.453	0.899	0.530	1.698
37	0.583	0.210	0.105	0.994	0.780	1.274
38	0.612	0.285	0.143	0.990	0.766	1.292
39	0.468	−0.080	−0.040	0.999	0.795	1.256
40	0.614	0.290	0.145	0.990	0.765	1.293
41	0.768	0.735	0.368	0.933	0.609	1.532
42	0.811	0.885	0.443	0.904	0.539	1.676
43	0.461	−0.095	−0.048	0.999	0.794	1.258
44	0.985	2.195	1.098	0.456	0.072	6.354
45	0.735	0.630	0.315	0.951	0.654	1.453
46	0.559	0.150	0.075	0.997	0.789	1.264
47	0.884	1.200	0.600	0.825	0.388	2.125
48	0.229	−0.740	−0.370	0.932	0.607	1.537
49	0.920	1.415	0.708	0.760	0.293	2.592
50	0.524	0.060	0.030	1.000	0.796	1.255

i	u_i	g_i	G_i	Cos G_i	p_i	(Cos G_i)/p_i
51	0.393	−0.270	−0.135	0.991	0.769	1.288
52	0.306	−0.505	−0.253	0.968	0.702	1.379
53	0.819	0.915	0.458	0.897	0.525	1.709
54	0.620	0.310	0.155	0.988	0.760	1.299
55	0.875	1.155	0.578	0.838	0.410	2.046
56	0.773	0.750	0.375	0.931	0.602	1.545
57	0.492	−0.020	−0.010	1.000	0.798	1.254
58	0.332	−0.430	−0.215	0.977	0.727	1.343
59	0.185	−0.890	−0.445	0.903	0.537	1.681
60	0.565	0.165	0.083	0.997	0.787	1.266
61	0.434	−0.165	−0.083	0.997	0.787	1.266
62	0.665	0.430	0.215	0.977	0.727	1.343
63	0.516	0.040	0.020	1.000	0.797	1.254
64	0.659	0.410	0.205	0.979	0.734	1.335
65	0.709	0.555	0.278	0.962	0.684	1.406
66	0.515	0.040	0.020	1.000	0.797	1.254
67	0.683	0.480	0.240	0.971	0.711	1.366
68	0.092	−1.320	−0.660	0.790	0.334	2.366
69	0.417	−0.205	−0.103	0.995	0.781	1.273
70	0.598	0.250	0.125	0.992	0.773	1.283
71	0.682	0.475	0.238	0.972	0.713	1.364
72	0.948	1.635	0.818	0.684	0.210	3.263
73	0.125	−1.145	−0.573	0.841	0.414	2.029
74	0.149	−1.035	−0.518	0.869	0.467	1.861
75	0.246	−0.685	−0.343	0.942	0.631	1.493

i	u_i	g_i	G_i	Cos G_i	p_i	(Cos G_i)/p_i
76	0.231	−0.730	−0.365	0.934	0.611	1.528
77	0.836	0.985	0.493	0.881	0.491	1.794
78	0.167	−0.960	−0.480	0.887	0.503	1.762
79	0.545	0.115	0.058	0.998	0.793	1.260
80	0.748	0.670	0.335	0.944	0.637	1.481
81	0.440	−0.150	−0.075	0.997	0.789	1.264
82	0.674	0.455	0.228	0.974	0.719	1.354
83	0.287	−0.560	−0.280	0.961	0.682	1.409
84	0.052	−1.615	−0.808	0.691	0.217	3.192
85	0.940	1.565	0.783	0.709	0.234	3.024
86	0.679	0.470	0.235	0.973	0.714	1.361
87	0.910	1.350	0.675	0.781	0.321	2.434
88	0.254	−0.660	−0.330	0.946	0.642	1.474
89	0.236	−0.715	−0.358	0.937	0.618	1.516
90	0.854	1.060	0.530	0.863	0.455	1.897
91	0.877	1.165	0.583	0.835	0.405	2.063
92	0.483	−0.040	−0.020	1.000	0.797	1.254
93	0.117	−1.185	−0.593	0.830	0.395	2.098
94	0.847	1.030	0.515	0.870	0.469	1.854
95	0.201	−0.835	−0.418	0.914	0.563	1.624
96	0.294	−0.540	−0.270	0.964	0.690	1.398
97	0.071	−1.460	−0.730	0.745	0.275	2.711
98	0.065	−1.505	−0.753	0.730	0.257	2.839
99	0.740	0.645	0.323	0.948	0.648	1.464
100	0.954	1.695	0.848	0.662	0.190	3.489
						Sum = 195.012
						I = 1.950

2.3   EVALUATION OF DEFINITE INTEGRALS USING THE MONTE CARLO METHOD: EXAMPLE III

As Example III, we now take up the integral

$$I = \int_a^b F(x)dx = \int_0^2 e^x dx, \tag{2.7}$$

where $F(x) = e^x$. We re-write Equation (2.7) as

$$I = \int_a^b \frac{F(x)}{p(x)} p(x)dx \tag{2.8}$$

which, as discussed in Section 1.7, implies that average value of F/p is the value of the integral, i.e.,

$$I = \frac{1}{N} \sum_{i=1}^N \frac{F(x_i)}{p(x_i)}, \tag{2.9}$$

where x_i's are random values of x in the interval $a < x < b$ obeying probability density function $p(x)$.

We now take up a linear variation for $p(x)$ given by $p(x) = Cx$ where C is a constant. Normalization of $p(x)$ requires $\int_0^2 p(x)dx = 1$ or, $C = 1/2$. As such, normalized probability density function is $p(x) = x/2$. See Figure 2.7 in which we have plotted $F(x) = e^x$ along with the normalized linear probability density function $p(x) = x/2$. The linear probability density function $p(x) = x/2$ follows the function e^x to some extent. Hence, it is not a bad choice.

We now generate random variable η obeying $p(x) = x/2$ in the interval $0 < x < 2$ using Equation (1.2):

$$\int_a^\eta p(x)dx = u,$$

where u is random variable (of Table 1.1) of uniform probability density function in the interval $0 < x < 1$. Thus, we have $\eta = 2u^{1/2}$.

Using $\eta = 2u^{1/2}$ in Microsoft Excel we have Table 2.8 in which we have evaluated the integral with the result 6.234 rather than 6.39. The difference is attributable to the fact that $F(x)$ and $p(x)$ used are not proportional to each other.

We now take up a normalized *Gaussian* variation for $p(x)$ given by

$$p(x) = \frac{1}{\sigma\sqrt{2\pi}} \mathrm{Exp}\left[-\frac{1}{2}\left(\frac{x-a}{\sigma}\right)^2\right],$$

where a is average and σ is variance of x. See Figure 2.8 in which we have plotted $F(x) = e^x$ along with the normalized Gaussian probability density function $p(x)$. As Figure 2.8 shows, the Gaussian probability density function $p(x)$ with $a = 1$ and $\sigma = 0.3$ is different from the

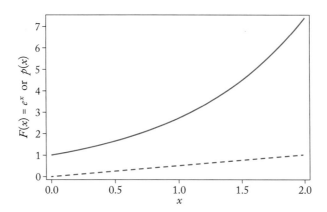

Figure 2.7: Showing $F(x) = e^x$ as undashed curve and $p(x) = x/2$ as dashed curve obtained using the command
```
Plot[{Exp[x],x/2},{x,0,2},Frame->True,FrameLabel->
{"x","F(x) = e^x or p(x)"},PlotStyle->{{Black},{Dashed, Black}}]
```
in Mathematica 6.0.

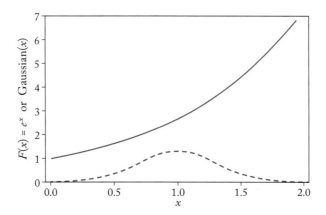

Figure 2.8: Showing $F(x) = e^x$ as undashed curve and Gaussian $p(x) = \frac{1}{\sigma\sqrt{2\pi}}$ $\mathrm{Exp}\left[-\frac{1}{2}\left(\frac{x-a}{\sigma}\right)^2\right]$ as dashed curve obtained using the command
```
sig=0.3; a=1;
p1=Plot[Exp[x],{x,0,2},PlotStyle->{Black},Frame->True,
FrameLabel->{"x","F(x) = e^x or Gaussian (x)"},PlotRange->{0,7}];
p2=Plot[(1/(sig*Sqrt[2*Pi]))*Exp[-0.5*((x-a)/sig)^2],
{x,0,7},PlotStyle->{Dashed,Black},PlotRange->{0,7}]; Show[p1,p2]
```
in Mathematica 6.0.

Table 2.8: Showing tabulated values of u_i, $\eta_i = 2u_i^{1/2}$, e^{η_i}/η_i from which we get $I = \frac{1}{N}\sum_{i=1}^{N}\frac{F(x_i)}{p(x_i)} = \frac{1}{N}\sum_{i=1}^{N}\frac{e^{\eta_i}}{\eta_i/2} = \frac{2}{N}\sum_{i=1}^{N}\frac{e^{\eta_i}}{\eta_i}$ as an approximate value of the integral $I = \int_a^b F(x)dx = \int_0^2 e^x dx = 6.234$ whereas the exact value is 6.389. Using $p(x) = x/2$.

i	u_i	$\eta_i = 2u_i^{1/2}$	e^{η_i}/η_i	i	u_i	$\eta_i = 2u_i^{1/2}$	e^{η_i}/η_i
1	0.169	0.822	2.768	26	0.225	0.949	2.722
2	0.050	0.447	3.497	27	0.687	1.658	3.165
3	0.844	1.837	3.418	28	0.060	0.490	3.332
4	0.182	0.853	2.751	29	0.070	0.529	3.208
5	0.599	1.548	3.037	30	0.835	1.828	3.403
6	0.985	1.985	3.667	31	0.466	1.365	2.869
7	0.726	1.704	3.225	32	0.271	1.041	2.721
8	0.148	0.769	2.805	33	0.813	1.803	3.366
9	0.788	1.775	3.325	34	0.376	1.226	2.780
10	0.698	1.671	3.182	35	0.983	1.983	3.663
11	0.010	0.200	6.107	36	0.181	0.851	2.752
12	0.397	1.260	2.798	37	0.583	1.527	3.015
13	0.852	1.846	3.432	38	0.612	1.565	3.056
14	0.761	1.745	3.281	39	0.468	1.368	2.871
15	0.309	1.112	2.734	40	0.614	1.567	3.058
16	0.113	0.672	2.913	41	0.768	1.753	3.292
17	0.212	0.921	2.727	42	0.811	1.801	3.363
18	0.809	1.799	3.359	43	0.461	1.358	2.863
19	0.084	0.580	3.080	44	0.985	1.985	3.667
20	0.276	1.051	2.722	45	0.735	1.715	3.240
21	0.243	0.986	2.719	46	0.559	1.495	2.983
22	0.989	1.989	3.674	47	0.884	1.880	3.487
23	0.508	1.425	2.918	48	0.229	0.957	2.721
24	0.817	1.808	3.373	49	0.920	1.918	3.550
25	0.174	0.834	2.761	50	0.524	1.448	2.938

i	u_i	$\eta_i = 2u_i^{1/2}$	e^{η_i}/η_i	i	u_i	$\eta_i = 2u_i^{1/2}$	e^{η_i}/η_i
51	0.393	1.254	2.794	76	0.231	0.961	2.720
52	0.306	1.106	2.733	77	0.836	1.829	3.404
53	0.819	1.810	3.376	78	0.167	0.817	2.771
54	0.620	1.575	3.067	79	0.545	1.476	2.965
55	0.875	1.871	3.471	80	0.748	1.730	3.260
56	0.773	1.758	3.300	81	0.440	1.327	2.841
57	0.492	1.403	2.899	82	0.674	1.642	3.146
58	0.332	1.152	2.747	83	0.287	1.071	2.725
59	0.185	0.860	2.748	84	0.052	0.456	3.460
60	0.565	1.503	2.991	85	0.940	1.939	3.585
61	0.434	1.318	2.834	86	0.679	1.648	3.153
62	0.665	1.631	3.132	87	0.910	1.908	3.532
63	0.516	1.437	2.928	88	0.254	1.008	2.718
64	0.659	1.624	3.123	89	0.236	0.972	2.719
65	0.709	1.684	3.199	90	0.854	1.848	3.435
66	0.515	1.435	2.927	91	0.877	1.873	3.474
67	0.683	1.653	3.159	92	0.483	1.390	2.888
68	0.092	0.607	3.024	93	0.117	0.684	2.897
69	0.417	1.292	2.817	94	0.847	1.841	3.423
70	0.598	1.547	3.036	95	0.201	0.897	2.734
71	0.682	1.652	3.158	96	0.294	1.084	2.727
72	0.948	1.947	3.600	97	0.071	0.533	3.197
73	0.125	0.707	2.868	98	0.065	0.510	3.266
74	0.149	0.772	2.803	99	0.740	1.720	3.247
75	0.246	0.992	2.718	100	0.954	1.953	3.611

Sum = 311.710

I = 6.234

function e^x in that it does not follow e^x for $x > 1$. Hence, it is not expected to yield a good estimate of the value of the definite integral. Other values of σ yield worse results.

We now generate random variable G obeying

$$p(G) = \frac{1}{\sigma\sqrt{2\pi}} \mathrm{Exp}\left[-\frac{1}{2}\left(\frac{G-a}{\sigma}\right)^2\right]$$

with $a = 1$ and $\sigma = 0.3$ using the tabulated values of random variable g of Table 1.4 obeying

$$p(g) = \frac{1}{\sigma\sqrt{2\pi}} \mathrm{Exp}\left[-\frac{1}{2}\left(\frac{g-a}{\sigma}\right)^2\right]$$

for $a = 0$ and $\sigma = 1$. The connection is $G = a + \sigma g = 1 + 0.3g$.

Using Table 2.9 in Microsoft Excel, we have evaluated the integral with the result 5.154 rather than the exact value 6.389. The difference is attributable to the fact that $F(x)$ and $p(x)$ used are not proportional to each other for $x > 1$.

2.4   EVALUATION OF DEFINITE INTEGRALS USING THE MONTE CARLO METHOD: EXAMPLE IV

As Example IV, we now take up the integral

$$I = \int_a^b F(x)dx = \int_1^5 \log_e x\, dx, \tag{2.10}$$

where $F(x) = \log_e x$. We re-write Equation (2.10) as

$$I = \int_a^b \frac{F(x)}{p(x)} p(x)dx \tag{2.11}$$

which, as discussed in Section 1.7, implies that average value of F/p is the value of the integral, i.e.,

$$I = \frac{1}{N} \sum_{i=1}^N \frac{F(x_i)}{p(x_i)}, \tag{2.12}$$

where x_i's are random values of x in the interval $a < x < b$ obeying probability density function $p(x)$.

We now take up an exponential variation for $p(x)$ given by $p(x) = Ce^x$ where C is a constant. Normalization of $p(x)$ requires $\int_1^5 p(x)dx = 1$ or, $C = 1/145$. As such, normalized probability density function is $p(x) = e^x/145$. See Figure 2.9 in which we have plotted $F(x) = \log_e x$ along with the normalized probability density function $p(x) = e^x/145$. The probability density function $p(x) = e^x/145$ follows the function $\log_e x$ well. Hence, it is a good choice.

Table 2.9: Showing tabulated values of u_i, g_i, G_i, e^{G_i}, p_i, e^{G_i}/p_i from which we get $I = \frac{1}{N} \sum_{i=1}^{N} \frac{F(x_i)}{p(x_i)} = \frac{1}{N} \sum_{i=1}^{N} \frac{e^{G_i}}{p_i} = \frac{1}{N} \sum_{i=1}^{N} \frac{e^{G_i}}{\frac{1}{\sigma\sqrt{2\pi}} \text{Exp}\left[-\frac{1}{2}\left(\frac{G_i-a}{\sigma}\right)^2\right]}$ as an approximate value of the integral $I = \int_a^b F(x)dx = \int_0^2 e^x dx = 5.154$ whereas the exact value is 6.389. Using $p(x) = \frac{1}{\sigma\sqrt{2\pi}} \text{Exp}\left[-\frac{1}{2}\left(\frac{x-a}{\sigma}\right)^2\right]$, $a = 1$ and $\sigma = 0.3$.

i	u_i	g_i	G_i	e^{G_i}	p_i	e^{G_i}/p_i
1	0.169	−0.955	0.714	2.041	0.843	2.422
2	0.050	−1.635	0.510	1.664	0.349	4.764
3	0.844	1.015	1.305	3.686	0.794	4.639
4	0.182	−0.905	0.729	2.072	0.883	2.347
5	0.599	0.255	1.077	2.934	1.287	2.280
6	0.985	2.195	1.659	5.251	0.120	43.925
7	0.726	0.605	1.182	3.259	1.107	2.943
8	0.148	−1.040	0.688	1.990	0.774	2.570
9	0.788	0.805	1.242	3.461	0.962	3.598
10	0.698	0.520	1.156	3.177	1.162	2.735
11	0.010	−2.295	0.312	1.365	0.096	14.296
12	0.397	−0.255	0.924	2.518	1.287	1.956
13	0.852	1.050	1.315	3.725	0.766	4.861
14	0.761	0.715	1.215	3.369	1.030	3.271
15	0.309	−0.495	0.852	2.343	1.176	1.992
16	0.113	−1.205	0.639	1.894	0.643	2.943
17	0.212	−0.795	0.762	2.141	0.969	2.209
18	0.809	0.880	1.264	3.540	0.903	3.920
19	0.084	−1.370	0.589	1.802	0.520	3.464
20	0.276	−0.590	0.823	2.277	1.117	2.038
21	0.243	−0.695	0.792	2.207	1.044	2.113
22	0.989	2.325	1.698	5.460	0.089	61.269
23	0.508	0.020	1.006	2.735	1.330	2.057
24	0.817	0.910	1.273	3.572	0.879	4.063
25	0.174	−0.935	0.720	2.053	0.859	2.391

i	u_i	g_i	G_i	e^{G_i}	p_i	e^{G_i}/p_i
26	0.225	−0.750	0.775	2.171	1.004	2.162
27	0.687	0.490	1.147	3.149	1.179	2.670
28	0.060	−1.540	0.538	1.713	0.406	4.215
29	0.070	−1.465	0.561	1.752	0.455	3.852
30	0.835	0.980	1.294	3.647	0.823	4.433
31	0.466	−0.085	0.975	2.650	1.325	2.000
32	0.271	−0.605	0.819	2.267	1.107	2.047
33	0.813	0.895	1.269	3.556	0.891	3.991
34	0.376	−0.315	0.906	2.473	1.265	1.954
35	0.983	2.145	1.644	5.173	0.133	38.822
36	0.181	−0.905	0.729	2.072	0.883	2.347
37	0.583	0.210	1.063	2.895	1.301	2.226
38	0.612	0.285	1.086	2.961	1.277	2.319
39	0.468	−0.080	0.976	2.654	1.326	2.002
40	0.614	0.290	1.087	2.965	1.275	2.326
41	0.768	0.735	1.221	3.389	1.015	3.339
42	0.811	0.885	1.266	3.545	0.899	3.944
43	0.461	−0.095	0.972	2.642	1.324	1.996
44	0.985	2.195	1.659	5.251	0.120	43.925
45	0.735	0.630	1.189	3.284	1.090	3.011
46	0.559	0.150	1.045	2.843	1.315	2.162
47	0.884	1.200	1.360	3.896	0.647	6.019
48	0.229	−0.740	0.778	2.177	1.011	2.153
49	0.920	1.415	1.425	4.156	0.489	8.504
50	0.524	0.060	1.018	2.768	1.327	2.085

i	u_i	g_i	G_i	e^{G_i}	p_i	e^{G_i}/p_i
51	0.393	−0.270	0.919	2.507	1.282	1.955
52	0.306	−0.505	0.849	2.336	1.171	1.996
53	0.819	0.915	1.275	3.577	0.875	4.088
54	0.620	0.310	1.093	2.983	1.267	2.354
55	0.875	1.155	1.347	3.844	0.683	5.632
56	0.773	0.750	1.225	3.404	1.004	3.391
57	0.492	−0.020	0.994	2.702	1.330	2.032
58	0.332	−0.430	0.871	2.389	1.212	1.971
59	0.185	−0.890	0.733	2.081	0.895	2.326
60	0.565	0.165	1.050	2.856	1.312	2.177
61	0.434	−0.165	0.951	2.587	1.312	1.972
62	0.665	0.430	1.129	3.093	1.212	2.551
63	0.516	0.040	1.012	2.751	1.329	2.070
64	0.659	0.410	1.123	3.074	1.223	2.514
65	0.709	0.555	1.167	3.211	1.140	2.816
66	0.515	0.040	1.012	2.751	1.329	2.070
67	0.683	0.480	1.144	3.139	1.185	2.649
68	0.092	−1.320	0.604	1.829	0.556	3.288
69	0.417	−0.205	0.939	2.556	1.302	1.963
70	0.598	0.250	1.075	2.930	1.289	2.273
71	0.682	0.475	1.143	3.135	1.188	2.639
72	0.948	1.635	1.491	4.439	0.349	12.706
73	0.125	−1.145	0.657	1.928	0.690	2.793
74	0.149	−1.035	0.690	1.993	0.778	2.560
75	0.246	−0.685	0.795	2.213	1.052	2.105

i	u_i	g_i	G_i	e^{G_i}	p_i	e^{G_i}/p_i
76	0.231	−0.730	0.781	2.184	1.019	2.143
77	0.836	0.985	1.296	3.653	0.819	4.462
78	0.167	−0.960	0.712	2.038	0.839	2.430
79	0.545	0.115	1.035	2.814	1.321	2.130
80	0.748	0.670	1.201	3.323	1.062	3.128
81	0.440	−0.150	0.955	2.599	1.315	1.976
82	0.674	0.455	1.137	3.116	1.199	2.599
83	0.287	−0.560	0.832	2.298	1.137	2.021
84	0.052	−1.615	0.516	1.674	0.361	4.639
85	0.940	1.565	1.470	4.347	0.391	11.124
86	0.679	0.470	1.141	3.130	1.191	2.629
87	0.910	1.350	1.405	4.076	0.535	7.623
88	0.254	−0.660	0.802	2.230	1.070	2.085
89	0.236	−0.715	0.786	2.194	1.030	2.130
90	0.854	1.060	1.318	3.736	0.758	4.927
91	0.877	1.165	1.350	3.855	0.675	5.715
92	0.483	−0.040	0.988	2.686	1.329	2.021
93	0.117	−1.185	0.645	1.905	0.659	2.891
94	0.847	1.030	1.309	3.702	0.782	4.732
95	0.201	−0.835	0.750	2.116	0.938	2.255
96	0.294	−0.540	0.838	2.312	1.149	2.011
97	0.071	−1.460	0.562	1.754	0.458	3.830
98	0.065	−1.505	0.549	1.731	0.428	4.039
99	0.740	0.645	1.194	3.299	1.080	3.054
100	0.954	1.695	1.509	4.520	0.316	14.296
						Sum = 515.351
						I = 5.154

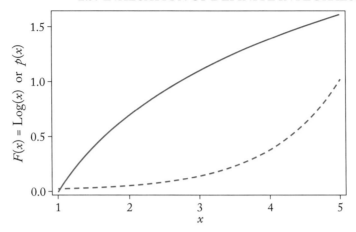

Figure 2.9: Showing $F(x) = \log_e x$ as undashed curve and $p(x) = e^x/145$ as dashed curve obtained using the command

```
a=1/145;
p1=Plot[Log[E,x],{x,1,5},Frame->True,
FrameLabel->{"x","F(x) = Log(x) or p(x)"},PlotStyle->{Black}];
p2=Plot[a*Exp[x],{x,1,5},PlotStyle->{Dashed,Black},Frame->True,
FrameLabel->{"x","F(x) = e^x or Gaussian (x)"}];Show[p1,p2]
```

in Mathematica 6.0.

We now generate random variable η obeying $p(x) = e^x/145$ in the interval $1 < x < 5$ using Equation (1.2):

$$\int_a^\eta p(x)dx = u \quad \text{or,} \quad \int_1^\eta p(x)dx = u,$$

where u is random variable (of Table 1.1) of uniform probability density function in the interval $0 < x < 1$. Thus, we have η given by $\eta = \log_e(145u + 2.72)$.

Using $\eta = \log_e(145u + 2.72)$ in Microsoft Excel we have Table 2.10 in which we have evaluated the integral with the result 3.915 rather than 4.05. The difference is attributable to the fact that $F(x)$ and $p(x)$ used are not proportional to each other.

2.5   EVALUATION OF DEFINITE INTEGRALS USING THE MONTE CARLO METHOD: EXAMPLE V

As Example V, we now take up the integral

$$I = \int_a^b F(x)dx = \int_0^3 \frac{1}{1+x^2}dx, \tag{2.13}$$

Table 2.10: Showing tabulated values of u_i, $\eta_i = \log_e(145u_i + 2.72)$, $\log_e \eta_i / e^{\eta i}$ from which we get $I = \frac{1}{N} \sum_{i=1}^{N} \frac{F(x_i)}{p(x_i)} = \frac{1}{N} \sum_{i=1}^{N} \frac{\log_e \eta_i}{e^{\eta_i}/145} = \frac{145}{N} \sum_{i=1}^{N} \frac{\log_e \eta_i}{e^{\eta_i}}$ as an approximate value of the integral $I = \int_a^b F(x)dx = \int_1^5 \log_e x\, dx = 3.915$ whereas the exact value is 4.045. Using $p(x) = e^x/145$.

i	u_i	$\eta_i = \text{Log}_e(145u_i + 2.72)$	$\dfrac{\text{Log}_e \eta_i}{e^{\eta_i}}$	i	u_i	$\eta_i = \text{Log}_e(145u_i + 2.72)$	$\dfrac{\text{Log}_e \eta_i}{e^{\eta_i}}$
1	0.169	3.237	0.046	26	0.225	3.513	0.037
2	0.050	2.109	0.091	27	0.687	4.608	0.015
3	0.844	4.812	0.013	28	0.060	2.271	0.085
4	0.182	3.308	0.044	29	0.070	2.410	0.079
5	0.599	4.473	0.017	30	0.835	4.802	0.013
6	0.985	4.965	0.011	31	0.466	4.225	0.021
7	0.726	4.663	0.015	32	0.271	3.694	0.032
8	0.148	3.110	0.051	33	0.813	4.775	0.013
9	0.788	4.744	0.014	34	0.376	4.014	0.025
10	0.698	4.624	0.015	35	0.983	4.963	0.011
11	0.010	0.896	−0.045	36	0.181	3.303	0.044
12	0.397	4.068	0.024	37	0.583	4.446	0.017
13	0.852	4.822	0.013	38	0.612	4.494	0.017
14	0.761	4.710	0.014	39	0.468	4.229	0.021
15	0.309	3.822	0.029	40	0.614	4.497	0.017
16	0.113	2.854	0.060	41	0.768	4.719	0.014
17	0.212	3.455	0.039	42	0.811	4.773	0.013
18	0.809	4.770	0.013	43	0.461	4.215	0.021
19	0.084	2.577	0.072	44	0.985	4.965	0.011
20	0.276	3.712	0.032	45	0.735	4.675	0.014
21	0.243	3.588	0.035	46	0.559	4.405	0.018
22	0.989	4.969	0.011	47	0.884	4.858	0.012
23	0.508	4.310	0.020	48	0.229	3.530	0.037
24	0.817	4.780	0.013	49	0.920	4.898	0.012
25	0.174	3.265	0.045	50	0.524	4.341	0.019

i	u_i	$\eta_i = \text{Log}_e\,(145u_i + 2.72)$	$\dfrac{\text{Log}_e\,\eta_i}{e^{\eta_i}}$	i	u_i	$\eta_i = \text{Log}_e\,(145u_i + 2.72)$	$\dfrac{\text{Log}_e\,\eta_i}{e^{\eta_i}}$
51	0.393	4.058	0.024	76	0.231	3.539	0.037
52	0.306	3.812	0.030	77	0.836	4.803	0.013
53	0.819	4.782	0.013	78	0.167	3.225	0.047
54	0.620	4.507	0.017	79	0.545	4.380	0.018
55	0.875	4.848	0.012	80	0.748	4.693	0.014
56	0.773	4.725	0.014	81	0.440	4.169	0.022
57	0.492	4.279	0.020	82	0.674	4.589	0.015
58	0.332	3.892	0.028	83	0.287	3.750	0.031
59	0.185	3.324	0.043	84	0.052	2.143	0.089
60	0.565	4.415	0.018	85	0.940	4.919	0.012
61	0.434	4.155	0.022	86	0.679	4.597	0.015
62	0.665	4.576	0.016	87	0.910	4.887	0.012
63	0.516	4.326	0.019	88	0.254	3.631	0.034
64	0.659	4.567	0.016	89	0.236	3.559	0.036
65	0.709	4.640	0.015	90	0.854	4.824	0.013
66	0.515	4.324	0.019	91	0.877	4.850	0.012
67	0.683	4.603	0.015	92	0.483	4.260	0.020
68	0.092	2.661	0.068	93	0.117	2.887	0.059
69	0.417	4.116	0.023	94	0.847	4.816	0.013
70	0.598	4.471	0.017	95	0.201	3.404	0.041
71	0.682	4.601	0.015	96	0.294	3.773	0.030
72	0.948	4.927	0.012	97	0.071	2.423	0.078
73	0.125	2.949	0.057	98	0.065	2.343	0.082
74	0.149	3.116	0.050	99	0.740	4.682	0.014
75	0.246	3.600	0.035	100	0.954	4.934	0.011

Sum = 2.700

I = 3.915

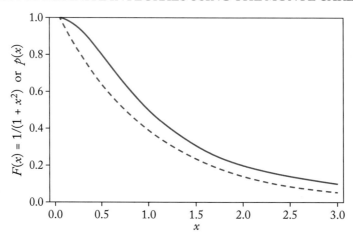

Figure 2.10: Showing $F(x) = \frac{1}{1+x^2}$ as undashed curve and $p(x) = 1.05e^{-x}$ as dashed curve obtained using the command
```
p1=Plot[1/(1+x^2),{x,0,3},PlotRange->{0,1},Frame->True, FrameLabel->
{"x","F(x) = 1/(1+x^2) or p(x)"},PlotStyle->{Black}];
p2=Plot[1.05*Exp[-x],{x,0,3},PlotStyle->{Dashed,Black}]; Show[p1,p2]
```
in Mathematica 6.0.

where $F(x) = \frac{1}{1+x^2}$. We re-write Equation (2.13) as

$$I = \int_a^b \frac{F(x)}{p(x)} p(x) dx \tag{2.14}$$

which, as discussed in Section 1.7, implies that average value of F/p is the value of the integral, i.e.,

$$I = \frac{1}{N} \sum_{i=1}^{N} \frac{F(x_i)}{p(x_i)}, \tag{2.15}$$

where x_i's are random values of x in the interval $a < x < b$ obeying probability density function $p(x)$.

We now take up an exponential variation for $p(x)$ given by $p(x) = Ce^{-x}$ where C is a constant. Normalization of $p(x)$ requires $\int_0^3 p(x) dx = 1$ or, $C = 1.05$. As such normalized probability density function is $p(x) = 1.05e^{-x}$. See Figure 2.10 in which we have plotted $F(x) = \frac{1}{1+x^2}$ along with the normalized probability density function $p(x) = 1.05e^{-x}$. The probability density function $p(x) = 1.05e^{-x}$ follows the function $\frac{1}{1+x^2}$ quite well. Hence, it is a very good choice.

We now generate random variable η obeying $p(x) = 1.05e^{-x}$ in the interval $0 < x < 3$ using Equation (1.2):

$$\int_a^\eta p(x)dx = u \quad \text{or,} \quad \int_0^\eta p(x)dx = u,$$

where u is random variable (of Table 1.1) of uniform probability density function in the interval $0 < x < 1$. Thus, we have η given by $\eta = -\log_e(1 - 0.95u)$.

Using $\eta = -\log_e(1 - 0.95u)$ in Microsoft Excel we have Table 2.11 in which we have evaluated the integral with the result 1.251 while the exact value is also 1.249. The absence of difference is attributable to the fact that $F(x)$ and $p(x)$ used are almost proportional to each other.

Table 2.11: Showing tabulated values of u_i, $\eta_i = -\log_e(1 - 0.95u_i)$, $e^{\eta i}/(1 + \eta_i^2)$ from which we get $I = \frac{1}{N}\sum_{i=1}^{N}\frac{F(x_i)}{p(x_i)} = \frac{1}{N}\sum_{i=1}^{N}\frac{1/(1+\eta_i^2)}{1.05e^{-\eta_i}} = \frac{0.95}{N}\sum_{i=1}^{N}\frac{e^{\eta_i}}{(1+\eta_i^2)}$ as an approximate value of the integral $I = \int_a^b F(x)dx = \int_0^3 \frac{1}{1+x^2}dx = 1.251$ whereas the exact value is 1.249. Using $p(x) = 1.05e^{-x}$.

i	u_i	$\eta_i = -\text{Log}_e(1 - 0.95u_i)$	$\dfrac{e^{\eta_i}}{(1 + \eta_i^2)}$	i	u_i	$\eta_i = -\text{Log}_e(1 - 0.95u_i)$	$\dfrac{e^{\eta_i}}{(1 + \eta_i^2)}$
1	0.169	0.175	1.156	26	0.225	0.240	1.202
2	0.050	0.049	1.047	27	0.687	1.057	1.359
3	0.844	1.617	1.394	28	0.060	0.059	1.057
4	0.182	0.190	1.167	29	0.070	0.069	1.066
5	0.599	0.841	1.358	30	0.835	1.575	1.388
6	0.985	2.743	1.822	31	0.466	0.584	1.337
7	0.726	1.169	1.360	32	0.271	0.297	1.237
8	0.148	0.151	1.137	33	0.813	1.479	1.377
9	0.788	1.380	1.369	34	0.376	0.442	1.301
10	0.698	1.087	1.359	35	0.983	2.714	1.804
11	0.010	0.010	1.009	36	0.181	0.189	1.166
12	0.397	0.473	1.311	37	0.583	0.807	1.357
13	0.852	1.657	1.400	38	0.612	0.870	1.359
14	0.761	1.283	1.363	39	0.468	0.588	1.338
15	0.309	0.347	1.263	40	0.614	0.875	1.359
16	0.113	0.113	1.106	41	0.768	1.307	1.364
17	0.212	0.225	1.192	42	0.811	1.471	1.376
18	0.809	1.462	1.375	43	0.461	0.576	1.336
19	0.084	0.083	1.079	44	0.985	2.743	1.822
20	0.276	0.304	1.241	45	0.735	1.197	1.361
21	0.243	0.262	1.216	46	0.559	0.757	1.355
22	0.989	2.804	1.863	47	0.884	1.830	1.433
23	0.508	0.659	1.348	48	0.229	0.245	1.205
24	0.817	1.496	1.379	49	0.920	2.070	1.500
25	0.174	0.181	1.160	50	0.524	0.688	1.351

i	u_i	$\eta_i = -\text{Log}_e\,(1-0.95u_i)$	$\dfrac{e^{\eta_i}}{(1+\eta_i{}^2)}$	i	u_i	$\eta_i = -\text{Log}_e\,(1-0.95u_i)$	$\dfrac{e^{\eta_i}}{(1+\eta_i{}^2)}$
51	0.393	0.467	1.310	76	0.231	0.248	1.207
52	0.306	0.343	1.261	77	0.836	1.580	1.389
53	0.819	1.504	1.379	78	0.167	0.173	1.154
54	0.620	0.889	1.359	79	0.545	0.729	1.354
55	0.875	1.778	1.422	80	0.748	1.239	1.362
56	0.773	1.325	1.365	81	0.440	0.541	1.329
57	0.492	0.630	1.344	82	0.674	1.022	1.359
58	0.332	0.379	1.277	83	0.287	0.318	1.248
59	0.185	0.193	1.169	84	0.052	0.051	1.049
60	0.565	0.769	1.356	85	0.940	2.234	1.558
61	0.434	0.531	1.327	86	0.679	1.035	1.359
62	0.665	0.998	1.359	87	0.910	1.998	1.477
63	0.516	0.673	1.349	88	0.254	0.276	1.225
64	0.659	0.983	1.359	89	0.236	0.254	1.211
65	0.709	1.119	1.359	90	0.854	1.667	1.401
66	0.515	0.671	1.349	91	0.877	1.790	1.425
67	0.683	1.046	1.359	92	0.483	0.614	1.342
68	0.092	0.091	1.087	93	0.117	0.118	1.110
69	0.417	0.504	1.320	94	0.847	1.632	1.396
70	0.598	0.839	1.358	95	0.201	0.212	1.183
71	0.682	1.043	1.359	96	0.294	0.327	1.253
72	0.948	2.307	1.589	97	0.071	0.070	1.067
73	0.125	0.126	1.117	98	0.065	0.064	1.061
74	0.149	0.153	1.138	99	0.740	1.213	1.361
75	0.246	0.266	1.219	100	0.954	2.366	1.615

Sum = 131.670

I = 1.251

CHAPTER 3

Variational Monte Carlo Method Applied to the Ground State of a Simple Harmonic Oscillator

3.1 THE VARIATIONAL METHOD OF QUANTUM MECHANICS APPLIED TO THE GROUND STATE OF ANY QUANTUM MECHANICAL SYSTEM

Let $\Psi_0, \Psi_1, \Psi_2, \Psi_3, \ldots, \Psi_N$ be exact, unknown, orthonormal eigenfunctions of known Hamiltonian H. Let $E_0 < E_1 < E_2 < E_3 < \ldots < E_N$ be exact, unknown eigenvalues of energy, respectively. That is, we have eigenvalue equation

$$H\Psi_n = E_n\Psi_n \tag{3.1}$$

which we cannot solve. Suppose we wish to estimate E_0 and we wish to know approximate form of Ψ_0. Let us choose *by guess* a suitable function Φ_0 that is expected to resemble unknown Ψ_0 and write

$$\Phi_0 = \sum_{n=0}^{N} C_n^{(0)}\Psi_n \tag{3.2}$$

and let us evaluate the quantity E_{Φ_0} given by

$$E_{\Phi_0} = \frac{(\Phi_0, H\Phi_0)}{(\Phi_0, \Phi_0)} = \frac{\int \Phi_0^* H \Phi_0 d\tau}{\int \Phi_0^* \Phi_0 d\tau} \tag{3.3}$$

$$= \frac{\left(\sum_{m=0} C_m^{(0)} \Psi_m, H \sum_{n=0} C_n^{(0)} \Psi_n\right)}{\left(\sum_{m=0} C_m^{(0)} \Psi_m, \sum_{n=0} C_n^{(0)} \Psi_n\right)}$$

$$= \frac{\left(\sum_{m=0} C_m^{(0)} \Psi_m, \sum_{n=0} C_n^{(0)} H \Psi_n\right)}{\left(\sum_{m=0} C_m^{(0)} \Psi_m, \sum_{n=0} C_n^{(0)} \Psi_n\right)}$$

$$= \frac{\left(\sum_{m=0} C_m^{(0)} \Psi_m, \sum_{n=0} C_n^{(0)} E_n \Psi_n\right)}{\left(\sum_{m=0} C_m^{(0)} \Psi_m, \sum_{n=0} C_n^{(0)} \Psi_n\right)}$$

using Equation (3.1)

$$= \frac{\sum_{m=0} \sum_{n=0} C_m^{(0)*} C_n^{(0)} E_n (\Psi_m, \Psi_n)}{\sum_{m=0} \sum_{n=0} C_m^{(0)*} C_n^{(0)} (\Psi_m, \Psi_n)}$$

$$= \frac{\sum_{m=0} \sum_{n=0} C_m^{(0)*} C_n^{(0)} E_n \delta_{mn}}{\sum_{m=0} \sum_{n=0} C_m^{(0)*} C_n^{(0)} \delta_{mn}} = \frac{\sum_{n=0} \left|C_n^{(0)}\right|^2 E_n}{\sum_{n=0} \left|C_n^{(0)}\right|^2}.$$

Thus,

$$E_{\Phi_0} - E_0 = \frac{\sum_{n=0} \left|C_n^{(0)}\right|^2 E_n}{\sum_{n=0} \left|C_n^{(0)}\right|^2} - E_0 = \frac{\sum_{n=0} \left|C_n^{(0)}\right|^2 (E_n - E_0)}{\sum_{n=0} \left|C_n^{(0)}\right|^2}. \tag{3.4}$$

In Equation (3.4), $E_n \geq E_0 . \therefore E_n - E_0 \geq 0$; moreover, $\left|C_n^{(0)}\right|^2 \geq 0$. Thus, Equation (3.4) shows that $E_{\Phi_0} - E_0 \geq 0$ or, $E_{\Phi_0} \geq E_0$. We find that for arbitrary Φ_0, calculated value of $E_{\Phi_0} \geq E_0$ i.e., E_{Φ_0} is larger than or equal to ground state energy E_0. E_{Φ_0} gives upper bound of ground state energy E_0.

If chosen Φ_0 is different from actual unknown Ψ_0, calculated value of E_{Φ_0} calculated using Equation (3.3) will be larger than E_0. If chosen Φ_0 happens to be same as unknown Ψ_0, value of E_{Φ_0} calculated using Equation (3.3) becomes equal to E_0. Thus, we can choose arbitrary Φ_0 and calculate E_{Φ_0} using Equation (3.3) and get upper bound of E_0.

Usually we choose a suitable Φ_0 known as trail function containing one or more adjustable parameters $\alpha_i^{(0)}$'s, e.g.,

$$\Phi_0 = e^{-\alpha_1^{(0)} x},$$

where $\alpha_1^{(0)}$ is an adjustable parameter. Then we use this expression of Φ_0 in Equation (3.3) and calculated expression of E_{Φ_0} will contain the parameters $\alpha_i^{(0)}$'s. We then stipulate $\frac{\partial E_{\Phi_0}}{\partial \alpha_i^{(0)}} = 0$ for each $\alpha_i^{(0)}$, separately. We thus get values of $\alpha_i^{(0)}$'s for which E_{Φ_0} is minimum. We can put these values of $\alpha_i^{(0)}$'s in the expression of E_{Φ_0}. We then get the lowest value of E_{Φ_0} for the chosen Φ_0. This upper bound to E_0 will be close to actual unknown value of E_0 if trail function Φ_0 has a form closely resembling that of actual unknown Ψ_0. The entire process is repeated for several different functions as Φ_0 to see which function leads to much lower value of E_{Φ_0}. If we are happy that a calculated value of E_{Φ_0} is low enough and that it is likely to be close to actual unknown value of E_0, we can take $E_0 \approx E_{\Phi_0}$ and $\Psi_0 \approx \Phi_0$ within some approximation.

3.2   THE VARIATIONAL METHOD OF QUANTUM MECHANICS APPLIED TO THE GROUND STATE OF A SIMPLE HARMONIC OSCILLATOR

Suppose we know Hamiltonian operator

$$H = -\frac{\hbar^2}{2m}\frac{d^2}{dx^2} + \frac{1}{2}m\omega_c^2 x^2$$

of a simple harmonic oscillator and we do not know eigenfunctions Ψ_n and eigenvalues E_n of H, because suppose we cannot solve eigenvalue equation $H\Psi_n = E_n\Psi_n$ of the operator H. Suppose we wish to estimate ground state energy E_0 and wish to know approximate form of corresponding eigenfunction Ψ_0.

Let us choose $\Phi_0 = e^{-\alpha x^2}$ by guessing as an approximate form of Ψ_0 and let us evaluate the quantity

$$E_{\Phi_0} = \frac{(\Phi_0, H\Phi_0)}{(\Phi_0, \Phi_0)}. \tag{3.5}$$

Now

$$(\Phi_0, \Phi_0) = \left(e^{-\alpha x^2}, e^{-\alpha x^2}\right) = \int_{-\infty}^{+\infty} e^{-\alpha x^2} e^{-\alpha x^2} dx = \int_{-\infty}^{+\infty} e^{-2\alpha x^2} dx$$

$$= \frac{1}{2}\int_{-\infty}^{+\infty} x^{-1} e^{-2\alpha x^2} 2x\, dx = \frac{1}{2}\int_{-\infty}^{+\infty} x^{-1} e^{-2\alpha x^2} dx^2 = \int_0^{+\infty} x^{-1} e^{-2\alpha x^2} dx^2$$

$$= \int_0^{+\infty} \left(x^2\right)^{-1/2} e^{-2\alpha x^2} dx^2 = \int_0^{+\infty} \left(x^2\right)^{1/2-1} e^{-2\alpha x^2} dx^2 = (2\alpha)^{-1/2}\Gamma(1/2)$$

$$= (2\alpha)^{-1/2}\sqrt{\pi} = \sqrt{\frac{\pi}{2\alpha}} \quad \text{using} \quad \int_0^{+\infty} x^{n-1} e^{-ax} dx = a^{-n}\Gamma n,$$

where Γ stands for a well-known Gamma function.

Thus,

$$(\Phi_0, \Phi_0) = \sqrt{\frac{\pi}{2\alpha}}. \tag{3.6}$$

We also note that

$$\int_0^{+\infty} e^{-2\alpha x^2} dx = \frac{1}{2}\sqrt{\frac{\pi}{2\alpha}}. \tag{3.7}$$

We now turn to $(\Phi_0, H\Phi_0)$. $H = T + V$ where T is operator of kinetic energy and V is operator of potential energy.

$$(\Phi_0, H\Phi_0) = (\Phi_0, (T + V)\Phi_0) = \left(\Phi_0, -\frac{\hbar^2}{2m}\frac{d^2\Phi_0}{dx^2} + \frac{1}{2}m\omega_c^2 x^2 \Phi_0\right). \tag{3.8}$$

Now

$$(\Phi_0, V\Phi_0) = \left(\Phi_0, \frac{1}{2}m\omega_c^2 x^2 \Phi_0\right) = \left(e^{-\alpha x^2}, \frac{1}{2}m\omega_c^2 x^2 e^{-\alpha x^2}\right)$$

$$= \frac{1}{2}m\omega_c^2\left(e^{-\alpha x^2}, x^2 e^{-\alpha x^2}\right) = \frac{1}{2}m\omega_c^2\int_{-\infty}^{+\infty} x^2 e^{-2\alpha x^2} dx = m\omega_c^2\int_0^{+\infty} x^2 e^{-2\alpha x^2} dx$$

$$= \frac{1}{2}m\omega_c^2\int_0^{+\infty} x e^{-2\alpha x^2} 2x\, dx = \frac{1}{2}m\omega_c^2\int_0^{+\infty} \left(x^2\right)^{1/2} e^{-2\alpha x^2} d\left(x^2\right)$$

$$= \frac{1}{2}m\omega_c^2\int_0^{+\infty} \left(x^2\right)^{3/2-1} e^{-2\alpha x^2} d\left(x^2\right) = \frac{1}{2}m\omega_c^2(2\alpha)^{-3/2}\Gamma(3/2)$$

$$= \frac{1}{2}m\omega_c^2(2\alpha)^{-3/2}(1/2)\Gamma(1/2) = \frac{1}{2}m\omega_c^2(2\alpha)^{-3/2}(1/2)\sqrt{\pi} = \frac{m\omega_c^2}{8\alpha}\sqrt{\frac{\pi}{2\alpha}},$$

using

$$\int_0^{+\infty} x^{n-1} e^{-ax} dx = a^{-n}\Gamma n \quad \text{and} \quad \Gamma(n+1) = n\Gamma n.$$

Thus,

$$(\Phi_0, V\Phi_0) = \frac{m\omega_c^2}{8\alpha}\sqrt{\frac{\pi}{2\alpha}}. \tag{3.9}$$

We also note that

$$\int_0^{+\infty} x^2 e^{-2\alpha x^2} dx = \frac{1}{8\alpha}\sqrt{\frac{\pi}{2\alpha}}. \tag{3.10}$$

Now

$$(\Phi_0, T\Phi_0) = \left(\Phi_0, -\frac{\hbar^2}{2m}\frac{d^2\Phi_0}{dx^2}\right)$$

$$= \left(e^{-\alpha x^2}, -\frac{\hbar^2}{2m}\frac{d^2}{dx^2}e^{-\alpha x^2}\right) = -\frac{\hbar^2}{2m}\left(e^{-\alpha x^2}, \frac{d^2}{dx^2}e^{-\alpha x^2}\right)$$

$$= -\frac{\hbar^2}{2m}\int_{-\infty}^{+\infty}e^{-\alpha x^2}\frac{d^2}{dx^2}e^{-\alpha x^2}\,dx = -\frac{\hbar^2}{m}\int_0^{+\infty}e^{-\alpha x^2}\frac{d^2}{dx^2}e^{-\alpha x^2}\,dx$$

$$= -\frac{\hbar^2}{m}\int_0^{+\infty}e^{-\alpha x^2}\frac{d}{dx}\left(-2\alpha x e^{-\alpha x^2}\right)dx$$

$$= -\frac{\hbar^2}{m}\int_0^{+\infty}e^{-\alpha x^2}\left(-2\alpha e^{-\alpha x^2} - 2\alpha x(-2\alpha x)e^{-\alpha x^2}\right)dx$$

$$= -\frac{\hbar^2}{m}\int_0^{+\infty}e^{-\alpha x^2}\left(-2\alpha e^{-\alpha x^2} + 4\alpha^2 x^2 e^{-\alpha x^2}\right)dx$$

$$= -\frac{\hbar^2}{m}\left[-2\alpha\int_0^{+\infty}e^{-2\alpha x^2}\,dx + 4\alpha^2\int_0^{+\infty}x^2 e^{-2\alpha x^2}\,dx\right]$$

$$= -\frac{\hbar^2}{m}\left[-2\alpha\frac{1}{2}\sqrt{\frac{\pi}{2\alpha}} + 4\alpha^2\frac{1}{8\alpha}\sqrt{\frac{\pi}{2\alpha}}\right] = -\frac{\hbar^2}{m}\left[-\alpha + \frac{\alpha}{2}\right]\sqrt{\frac{\pi}{2\alpha}} = \frac{\hbar^2\alpha}{2m}\sqrt{\frac{\pi}{2\alpha}},$$

using Equations (3.7) and (3.10). Thus,

$$(\Phi_0, T\Phi_0) = \frac{\hbar^2\alpha}{2m}\sqrt{\frac{\pi}{2\alpha}}. \tag{3.11}$$

From Equations (3.11) and (3.9), we gather that

$$(\Phi_0, H\Phi_0) = (\Phi_0, (T+V)\Phi_0) = \left(\frac{\hbar^2\alpha}{2m} + \frac{m\omega_c^2}{8\alpha}\right)\sqrt{\frac{\pi}{2\alpha}}. \tag{3.12}$$

Using Equations (3.12) and (3.6) in Equation (3.5), we get

$$E_{\Phi_0} = \frac{(\Phi_0, H\Phi_0)}{(\Phi_0, \Phi_0)} = \frac{\left(\frac{\hbar^2\alpha}{2m} + \frac{m\omega_c^2}{8\alpha}\right)\sqrt{\frac{\pi}{2\alpha}}}{\sqrt{\frac{\pi}{2\alpha}}} = \frac{\hbar^2\alpha}{2m} + \frac{m\omega_c^2}{8\alpha}. \tag{3.13}$$

We now stipulate that $\frac{\partial E_{\Phi_0}}{\partial\alpha} = 0$. Hence, $\frac{\hbar^2}{2m} - \frac{m\omega_c^2}{8\alpha^2} = 0$. Thus, we get lowest value of E_{Φ_0} for

$$\alpha = \frac{m\omega_c}{2\hbar} \tag{3.14}$$

for the chosen $\Phi_0(= e^{-\alpha x^2})$. Using Equation (3.14) in Equation (3.13), we get the lowest value of

$$E_{\Phi_0} = \frac{\hbar^2 \alpha}{2m} + \frac{m\omega_c^2}{8\alpha} = \frac{\hbar^2}{2m}\frac{m\omega_c}{2\hbar} + \frac{m\omega_c^2}{8\frac{m\omega_c}{2\hbar}}.$$

We take ground state energy as

$$E_0 \approx E_{\Phi_0} = \frac{1}{2}\hbar\omega_c. \tag{3.15}$$

This incidentally agrees exactly with known value of ground state energy E_0 of simple harmonic oscillator. The ground state eigenfunction $\Phi_0 = e^{-\alpha x^2}$ becomes

$$\Psi_0 \approx \Phi_0 = e^{-\frac{1}{2}\frac{m\omega_c}{\hbar}x^2} \tag{3.16}$$

which incidentally agrees exactly with known ground state eigenfunction Ψ_0 of energy of simple harmonic oscillator. Thus, to some extent, we have become familiar with use of the variational method.

3.3 GROUND STATE ENERGY OF A SIMPLE HARMONIC OSCILLATOR USING THE MONTE CARLO METHOD

We consider an electron residing in a simple harmonic oscillator potential

$$V(x) = \frac{1}{2}m\omega_c^2 x^2 = \frac{1}{2}kx^2. \tag{3.17}$$

We first gather exact results from Quantum Mechanics books. Eigenvalue equation of energy is $H\Psi = E\Psi$, or

$$\left[-\frac{\hbar^2}{2m}\frac{d^2}{dx^2} + \frac{1}{2}m\omega_c^2 x^2 \right]\Psi = E\Psi. \tag{3.18}$$

Eigenvalues of energy are

$$E_n = (n + 1/2)\hbar\omega_c, \tag{3.19}$$

where $n = 0, 1, 2, 3, \ldots$. Ground state energy in unit of $\hbar\omega_c$ is $E_0/(\hbar\omega_c) = 0.5$ and ground state eigenfunction is

$$\Psi_0 = e^{-\frac{1}{2}\frac{m\omega_c}{\hbar}x^2} = e^{-\frac{1}{2}Lx^2}, \tag{3.20}$$

where

$$L = \frac{m\omega_c}{\hbar}. \tag{3.21}$$

We now gather results from variational method of Quantum Mechanics detailed in Section 3.2: for trial wavefunction

$$\Phi_0 = e^{-\alpha x^2} = e^{-\beta Lx^2},$$

ground state energy estimated or evaluated using Equation (3.5):

$$E_{\Phi_0} = \frac{(\Phi_0, H\Phi_0)}{(\Phi_0, \Phi_0)}$$

is $E_0/(\hbar\omega_c) = 0.5$ and value of β turns out to be 0.5 so that ground state eigenfunction is

$$\Phi_0 = e^{-\alpha x^2} = e^{-\beta L x^2} = e^{-\frac{1}{2} L x^2}. \tag{3.22}$$

We now move onto *variational Monte Carlo method* of evaluating ground state eigenvalue and eigenfunction of energy. We have Hamiltonian operator

$$H = -\frac{\hbar^2}{2m}\frac{d^2}{dx^2} + \frac{1}{2}m\omega_c^2 x^2 \quad \text{or,} \quad H = -\frac{\hbar^2}{2m}\frac{d^2}{dx^2} + \frac{1}{2}kx^2$$

which together with trial wavefunction $\Phi_0 = e^{-\alpha x^2}$ lead to

$$H\Phi_0 = He^{-\alpha x^2} = -\frac{\hbar^2}{2m}\left(-2\alpha + 4\alpha^2 x^2\right)e^{-\alpha x^2} + \frac{1}{2}kx^2 e^{-\alpha x^2}$$

or,

$$\frac{H\Phi_0}{\Phi_0} = -\frac{\hbar^2}{2m}\left(-2\alpha + 4\alpha^2 x^2\right) + \frac{1}{2}kx^2 \tag{3.23}$$

and ground state energy in unit of $\hbar\omega_c$ as

$$\frac{E_0}{\hbar\omega_c} = \frac{(\Phi_0, H\Phi_0)}{\hbar\omega_c(\Phi_0, \Phi_0)} = \frac{1}{\hbar\omega_c\sqrt{\frac{\pi}{2\alpha}}}\int \frac{H\Phi_0}{\Phi_0}\Phi_0^2 dx$$

$$= \int F(x)dx = \int f(x)p(x)dx$$

$$= <f> = <\frac{1}{\hbar\omega_c}\frac{H\Phi_0}{\Phi_0}> . \tag{3.24}$$

Thus, *average* or the so-called *expectation value* of $\frac{1}{\hbar\omega_c}\frac{H\Phi_0}{\Phi_0}$ is a good estimate of ground state energy in unit of $\hbar\omega_c$. Here,

$$f = \frac{1}{\hbar\omega_c}\frac{H\Phi_0}{\Phi_0} \tag{3.25}$$

$$p(x) = \sqrt{\frac{2\alpha}{\pi}}\Phi_0^2 \tag{3.26}$$

and

$$F(x) = \left(\frac{1}{\hbar\omega_c}\frac{H\Phi_0}{\Phi_0}\right)\left(\sqrt{\frac{2\alpha}{\pi}}\Phi_0^2\right). \tag{3.27}$$

Looking at Equation (3.24), ground state energy in unit of $\hbar\omega_c$ is

$$\frac{E_0}{\hbar\omega_c} =< \frac{1}{\hbar\omega_c}\frac{H\Phi_0}{\Phi_0} >= \frac{1}{\hbar\omega_c} < -\frac{\hbar^2}{2m}\left(-2\alpha + 4\alpha^2 x^2\right) + \frac{1}{2}kx^2 >$$

$$=< -\frac{\hbar}{2m\omega_c}\left(-2\alpha + 4\alpha^2 x^2\right) + \frac{1}{2}\frac{k}{\hbar\omega_c}x^2 >$$

using Equation (3.23), or

$$\frac{E_0}{\hbar\omega_c} =< -\frac{1}{2L}\left(-2\alpha + 4\alpha^2 x^2\right) + \frac{L}{2}x^2 > \tag{3.28}$$

using $\frac{k}{\hbar\omega_c} = L$ obtained as follows:

$$L^2 = \frac{m^2\omega_c^2}{\hbar^2},$$

see Equation (3.21), or

$$L^2 = \frac{m^2}{\hbar^2}\frac{k}{m},$$

see Equation (3.17), or

$$L = \frac{\sqrt{mk}}{\hbar}.$$

Again,

$$\frac{k}{\hbar\omega_c} = \frac{k}{\hbar}\sqrt{\frac{m}{k}} = \frac{\sqrt{mk}}{\hbar} = L.$$

In Equation (3.28), $\alpha = \beta L$ and $L = \frac{m\omega_c}{\hbar}$. We note that ground state energy is given by

$$\frac{E_0}{\hbar\omega_c} =< -\frac{1}{2L}\left(-2\alpha + 4\alpha^2 x_i^2\right) + \frac{L}{2}x_i^2 >, \tag{3.29}$$

where x_i's are random values of x obeying the Gaussian probability density function $p(x)$ given by Equation (3.26).

Before we proceed to evaluate ground state energy using the prescription of Equation (3.29), we present expressions for $F(x)$ and $p(x)$ and plot them together to ascertain that they follow each other. As demonstrated in Section 1.7, $p(x)$ at least should follow $F(x)$, if not be proportional to $F(x)$ to get good result.

According to Equation (3.26),

$$p(x) = \sqrt{\frac{2\alpha}{\pi}}e^{-2\alpha x^2} \tag{3.30}$$

which is Gaussian function of x given by

$$p(x) = \frac{1}{\sigma\sqrt{2\pi}}e^{-\frac{1}{2}\left(\frac{x-a}{\sigma}\right)^2}. \tag{3.31}$$

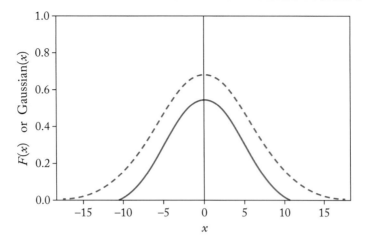

Figure 3.1: Showing $F(x) = \left(-\frac{1}{2L}(-2\alpha + 4\alpha^2 x^2) + \frac{L}{2}x^2\right)\left(\sqrt{\frac{2\alpha}{\pi}}e^{-2\alpha x^2}\right)$ of Equation (3.33) as undashed curve and Gaussian $p(x) = \frac{1}{\sigma\sqrt{2\pi}}\text{Exp}\left[-\frac{1}{2}\left(\frac{x-a}{\sigma}\right)^2\right]$ of Equation (3.31) as dashed curve obtained using Program 3.1. For chosen value of spring constant $k = 10^{-18}$ that corresponds to $\omega_c = 1.048$ MHz, $L = \frac{m\omega_c}{\hbar} = 9.046 \times 10^9$, $\beta = 0.8$, $\alpha = \beta L = 7.237 \times 10^9$, $a = 0$, $\sigma = \frac{1}{2}\alpha^{-1/2} = 5.878 \times 10^{-6}$. (Here x is in unit of 10^{-6} and $F(x)$ and $p(x)$ are in unit of 10^{-5}).

Comparison between Equations (3.30) and (3.31) shows that $a = 0$ and

$$\sigma = \frac{1}{2}\alpha^{-1/2}. \tag{3.32}$$

And, according to Equation (3.27),

$$F(x) = \left(\frac{1}{\hbar\omega_c}\frac{H\Phi_0}{\Phi_0}\right)\left(\sqrt{\frac{2\alpha}{\pi}}\Phi_0^2\right)$$

or

$$F(x) = \left(-\frac{1}{2L}(-2\alpha + 4\alpha^2 x^2) + \frac{L}{2}x^2\right)\left(\sqrt{\frac{2\alpha}{\pi}}e^{-2\alpha x^2}\right) \tag{3.33}$$

using Equation (3.28).

Using Program 3.1 in Mathematica 6.0, we plotted Figure 3.1 to ascertain that $F(x)$ and $p(x)$ follow each other. We have chosen "spring constant" $k = 10^{-18}$ that corresponds to $\omega_c = 1.048$ MHz. We find that $F(x)$ and $p(x)$ follow each other very well and, hence, Equation (3.29) will give a good estimate of $\frac{E_0}{\hbar\omega_c}$.

Program 3.1

```
beta=0.8;

k=10^-18;
m=9.1*10^-31;
hcut=(6.626*10^-34)/(2*3.1416);

omega=Sqrt[k/m]
L=m*omega/hcut
alpha=beta*L;

a=0;
sig=0.5*((beta*L)^(-0.5))

F=Plot[(10^-5)*((-1/(2*L))*(-2*alpha+4*alpha^2*(x*(10^-6))^2)
+0.5*L*(x*(10^-6))^2)*(Sqrt[2*alpha/3.1416])*
Exp[-2*alpha*(x*(10^-6))^2], {x,-3*sig/(10^-6),
+3*sig/(10^-6)},PlotStyle->{Black},Frame->True,
FrameLabel->{"x","F(x) or Gaussian (x)"},PlotRange->{0,1}];

p=Plot[(10^-5)*(Sqrt[2*alpha/3.1416])*Exp[-2*alpha*(x*(10^-6))^2],
{x,-3*sig/(10^-6),+3*sig/(10^-6)},PlotStyle->{Dashed,Black},
PlotRange->{0,1}];
Show[F,p]
```

We now generate random variables G obeying

$$p(G) = \frac{1}{\sigma\sqrt{2\pi}} \mathrm{Exp}\left[-\frac{1}{2}\left(\frac{G-a}{\sigma}\right)^2\right]$$

with $a = 0$ and $\sigma = 5.878 \times 10^{-6}$ using the tabulated values of random variable g of Table 1.4 obeying

$$p(g) = \frac{1}{\sigma\sqrt{2\pi}} \mathrm{Exp}\left[-\frac{1}{2}\left(\frac{g-a}{\sigma}\right)^2\right]$$

for $a = 0$ and $\sigma = 1$. The connection is $G = a + \sigma g = 0 + 5.878 \times 10^{-6}g$.

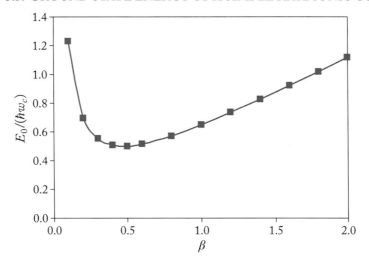

Figure 3.2: Showing values of ground state energy $E_0/(\hbar\omega_c)$ of an electron in simple harmonic oscillator potential $V(x) = \frac{1}{2}kx^2$ with $k = 10^{-18}$, for various values of variational parameter β contained in trial wavefunction $\Phi_0 = e^{-\beta L x^2} = e^{-\beta \frac{m\omega_c}{\hbar} x^2}$. Minimum value of $E_0/(\hbar\omega_c) = 0.5$ occurs for $\beta = 0.5$. Thus, the ground state eigenfunction is $\Phi_0 = e^{-\beta L x^2} = e^{-\frac{1}{2}L x^2} = e^{-\frac{1}{2}\frac{m\omega_c}{\hbar} x^2}$ while ground state eigenvalue of energy is $E_0 = 0.5\hbar\omega_c$.

Using Table 3.1 in Microsoft Excel, we have evaluated the expectation value set forth in Equation (3.29)

$$\frac{E_0}{\hbar\omega_c} = < -\frac{1}{2L}\left(-2\alpha + 4\alpha^2 x_i^2\right) + \frac{L}{2}x_i^2 >$$

as a value of the ground state energy.

We have run Program 3.1 and executed the Microsoft Excel spreadsheet for Table 3.1 for various values of β namely 0.1–2.0 and obtained values of $E_0/(\hbar\omega_c)$ in each case. Table 3.2 shows the results which are plotted in Figure 3.2. We find from Figure 3.2 that minimum value of $E_0/(\hbar\omega_c) = 0.5$ occurs for $\beta = 0.5$. Thus, the ground state eigenfunction is $\Phi_0 = e^{-\beta L x^2} = e^{-\frac{1}{2}L x^2} = e^{-\frac{1}{2}\frac{m\omega_c}{\hbar} x^2}$ while ground state eigenvalue of energy is $E_0 = 0.5\hbar\omega_c$.

We have repeated the run with Program 3.1 with $k = 10^{-16}$. Figure 3.3 shows that $F(x)$ and $p(x)$ follow each other very well and hence good results are expected. We use $\beta = 0.8$, $\omega_c = 10.483 \times 10^6$, $L = 9.046 \times 10^{10}$, $\alpha = \beta L = 7.237 \times 10^{10}$, $a = 0$, $\sigma = 1.859 \times 10^{-6}$.

We now generate random variables G obeying

$$p(G) = \frac{1}{\sigma\sqrt{2\pi}}\text{Exp}\left[-\frac{1}{2}\left(\frac{G-a}{\sigma}\right)^2\right]$$

Table 3.1: Showing tabulated values of u_i, g_i, G_i, $-\frac{1}{2L}\left(-2\alpha + 4\alpha^2 G_i^2\right) + \frac{L}{2}G_i^2$ from which we get $\frac{E_0}{\hbar\omega_c} = < -\frac{1}{2L}\left(-2\alpha + 4\alpha^2 G_i^2\right) + \frac{L}{2}G_i^2 >$. Using $a = 0, \sigma = 5.878 \times 10^{-6}$, $\beta = 0.8$, $L = 9.046 \times 10^9$, $\alpha = \beta L$.

i	u_i	g_i	$G_i = a + \sigma g_i$	$-\frac{1}{2L}(-2\alpha + 4\alpha^2 G_i^2) + \frac{L}{2}G_i^2$
1	0.169	−0.955	−5.613E−06	5.777E−01
2	0.050	−1.635	−9.611E−06	1.483E−01
3	0.844	1.015	5.966E−06	5.488E−01
4	0.182	−0.905	−5.320E−06	6.003E−01
5	0.599	0.255	1.499E−06	7.841E−01
6	0.985	2.195	1.290E−05	−3.746E−01
7	0.726	0.605	3.556E−06	7.108E−01
8	0.148	−1.040	−6.113E−06	5.363E−01
9	0.788	0.805	4.732E−06	6.420E−01
10	0.698	0.520	3.057E−06	7.341E−01
11	0.010	−2.295	−1.349E−05	−4.840E−01
12	0.397	−0.255	−1.499E−06	7.841E−01
13	0.852	1.050	6.172E−06	5.312E−01
14	0.761	0.715	4.203E−06	6.754E−01
15	0.309	−0.495	−2.910E−06	7.403E−01
16	0.113	−1.205	−7.083E−06	4.460E−01
17	0.212	−0.795	−4.673E−06	6.459E−01
18	0.809	0.880	5.173E−06	6.112E−01
19	0.084	−1.370	−8.053E−06	3.424E−01
20	0.276	−0.590	−3.468E−06	7.151E−01
21	0.243	−0.695	−4.085E−06	6.822E−01
22	0.989	2.325	1.367E−05	−5.178E−01
23	0.508	0.020	1.176E−07	7.999E−01
24	0.817	0.910	5.349E−06	5.981E−01
25	0.174	−0.935	−5.496E−06	5.869E−01

i	u_i	g_i	$G_i = a + \sigma g_i$	$-\dfrac{1}{2L}(-2a + 4a^2 G_i^2) + \dfrac{L}{2}G_i^2$
26	0.225	−0.750	−4.409E−06	6.629E−01
27	0.687	0.490	2.880E−06	7.415E−01
28	0.060	−1.540	−9.052E−06	2.218E−01
29	0.070	−1.465	−8.611E−06	2.768E−01
30	0.835	0.980	5.760E−06	5.659E−01
31	0.466	−0.085	−4.996E−07	7.982E−01
32	0.271	−0.605	−3.556E−06	7.108E−01
33	0.813	0.895	5.261E−06	6.047E−01
34	0.376	−0.315	−1.852E−06	7.758E−01
35	0.983	2.145	1.261E−05	−3.217E−01
36	0.181	−0.905	−5.320E−06	6.003E−01
37	0.583	0.210	1.234E−06	7.892E−01
38	0.612	0.285	1.675E−06	7.802E−01
39	0.468	−0.080	−4.702E−07	7.984E−01
40	0.614	0.290	1.705E−06	7.795E−01
41	0.768	0.735	4.320E−06	6.683E−01
42	0.811	0.885	5.202E−06	6.091E−01
43	0.461	−0.095	−5.584E−07	7.978E−01
44	0.985	2.195	1.290E−05	−3.746E−01
45	0.735	0.630	3.703E−06	7.032E−01
46	0.559	0.150	8.817E−07	7.945E−01
47	0.884	1.200	7.054E−06	4.489E−01
48	0.229	−0.740	−4.350E−06	6.665E−01
49	0.920	1.415	8.317E−06	3.119E−01
50	0.524	0.060	3.527E−07	7.991E−01

i	u_i	g_i	$G_i = a + \sigma g_i$	$-\dfrac{1}{2L}(-2a + 4a^2 G_i{}^2) + \dfrac{L}{2} G_i{}^2$
51	0.393	−0.270	−1.587E−06	7.822E−01
52	0.306	−0.505	−2.968E−06	7.378E−01
53	0.819	0.915	5.378E−06	5.959E−01
54	0.620	0.310	1.822E−06	7.766E−01
55	0.875	1.155	6.789E−06	4.748E−01
56	0.773	0.750	4.409E−06	6.629E−01
57	0.492	−0.020	−1.176E−07	7.999E−01
58	0.332	−0.430	−2.528E−06	7.549E−01
59	0.185	−0.890	−5.231E−06	6.069E−01
60	0.565	0.165	9.699E−07	7.934E−01
61	0.434	−0.165	−9.699E−07	7.934E−01
62	0.665	0.430	2.528E−06	7.549E−01
63	0.516	0.040	2.351E−07	7.996E−01
64	0.659	0.410	2.410E−06	7.590E−01
65	0.709	0.555	3.262E−06	7.249E−01
66	0.515	0.040	2.351E−07	7.996E−01
67	0.683	0.480	2.821E−06	7.438E−01
68	0.092	−1.320	−7.759E−06	3.752E−01
69	0.417	−0.205	−1.205E−06	7.898E−01
70	0.598	0.250	1.470E−06	7.848E−01
71	0.682	0.475	2.792E−06	7.450E−01
72	0.948	1.635	9.611E−06	1.483E−01
73	0.125	−1.145	−6.730E−06	4.804E−01
74	0.149	−1.035	−6.084E−06	5.388E−01
75	0.246	−0.685	−4.026E−06	6.856E−01

i	u_i	g_i	$G_i = a + \sigma\, g_i$	$-\dfrac{1}{2L}(-2a + 4a^2 G_i^2) + \dfrac{L}{2} G_i^2$
76	0.231	−0.730	−4.291E−06	6.701E−01
77	0.836	0.985	5.790E−06	5.635E−01
78	0.167	−0.960	−5.643E−06	5.753E−01
79	0.545	0.115	6.760E−07	7.968E−01
80	0.748	0.670	3.938E−06	6.906E−01
81	0.440	−0.150	−8.817E−07	7.945E−01
82	0.674	0.455	2.674E−06	7.495E−01
83	0.287	−0.560	−3.292E−06	7.235E−01
84	0.052	−1.615	−9.493E−06	1.641E−01
85	0.940	1.565	9.199E−06	2.029E−01
86	0.679	0.470	2.763E−06	7.461E−01
87	0.910	1.350	7.935E−06	3.557E−01
88	0.254	−0.660	−3.879E−06	6.938E−01
89	0.236	−0.715	−4.203E−06	6.754E−01
90	0.854	1.060	6.231E−06	5.261E−01
91	0.877	1.165	6.848E−06	4.691E−01
92	0.483	−0.040	−2.351E−07	7.996E−01
93	0.117	−1.185	−6.965E−06	4.577E−01
94	0.847	1.030	6.054E−06	5.414E−01
95	0.201	−0.835	−4.908E−06	6.300E−01
96	0.294	−0.540	−3.174E−06	7.289E−01
97	0.071	−1.460	−8.582E−06	2.803E−01
98	0.065	−1.505	−8.846E−06	2.478E−01
99	0.740	0.645	3.791E−06	6.986E−01
100	0.954	1.695	9.963E−06	9.959E−02
				Sum = 57.064
				$E_0/(\hbar w_c) = 0.571$

Table 3.2: Showing values of ground state energy $E_0/(\hbar\omega_c)$ of an electron in simple harmonic oscillator potential $V(x) = \frac{1}{2}kx^2$ with $k = 10^{-18}$, for various values of variational parameter β contained in trial wave function $\Phi_0 = e^{-\beta L x^2} = e^{-\beta \frac{m\omega_c}{\hbar} x^2}$

β	$E_0/(\hbar w_c)$
0.1	1.229
0.2	0.694
0.3	0.551
0.4	0.506
0.5	0.500
0.6	0.514
0.8	0.571
1.0	0.647
1.2	0.734
1.4	0.825
1.6	0.921
1.8	1.019
2.0	1.118

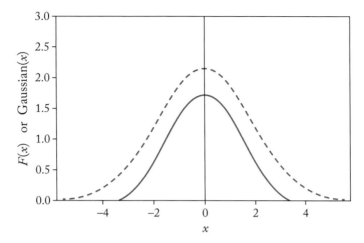

Figure 3.3: Showing $F(x) = \left(-\frac{1}{2L}\left(-2\alpha + 4\alpha^2 x^2\right) + \frac{L}{2}x^2\right)\left(\sqrt{\frac{2\alpha}{\pi}}e^{-2\alpha x^2}\right)$ of Equation (3.33) as undashed curve and Gaussian $p(x) = \frac{1}{\sigma\sqrt{2\pi}}\text{Exp}\left[-\frac{1}{2}\left(\frac{x-a}{\sigma}\right)^2\right]$ of Equation (3.31) as dashed curve obtained using Program 3.1. For chosen value of spring constant $k = 10^{-16}$ that corresponds to $\omega_c = 10.483$ MHz, $L = \frac{m\omega_c}{\hbar} = 9.046 \times 10^{10}$, $\beta = 0.8$, $\alpha = \beta L = 7.237 \times 10^{10}$, $a = 0$, $\sigma = \frac{1}{2}\alpha^{-1/2} = 1.859 \times 10^{-6}$. (Here x is in unit of 10^{-6} and $F(x)$ and $p(x)$ are in unit of 10^{-5}.)

with $a = 0$ and $\sigma = 1.859 \times 10^{-6}$ using the tabulated values of random variable g of Table 1.4 obeying

$$p(g) = \frac{1}{\sigma\sqrt{2\pi}}\text{Exp}\left[-\frac{1}{2}\left(\frac{g-a}{\sigma}\right)^2\right]$$

for $a = 0$ and $\sigma = 1$. The connection is $G = a + \sigma g = 0 + 1.859 \times 10^{-6}g$.

Using Table 3.3 in Microsoft Excel, we have evaluated the expectation value set forth in Equation (3.29)

$$\frac{E_0}{\hbar\omega_c} =< -\frac{1}{2L}\left(-2\alpha + 4\alpha^2 x_i^2\right) + \frac{L}{2}x_i^2 >$$

as a value of the ground state energy.

We have run Program 3.1 and executed the Microsoft Excel spreadsheet for Table 3.3 for various values of β namely 0.1–2.0 and obtained values of $E_0/(\hbar\omega_c)$ in each case. Table 3.4 shows the results which are plotted in Figure 3.4. We find from Figure 3.4 that minimum value of $E_0/(\hbar\omega_c) = 0.5$ occurs for $\beta = 0.5$. Thus, the ground state eigenfunction is

$$\Phi_0 = e^{-\beta L x^2} = e^{-\frac{1}{2}L x^2} = e^{-\frac{1}{2}\frac{m\omega_c}{\hbar}x^2}$$

while ground state eigenvalue of energy is $E_0 = 0.5\hbar\omega_c$.

Table 3.3: Showing tabulated values of u_i, g_i, G_i, $-\frac{1}{2L}(-2\alpha + 4\alpha^2 G_i^2) + \frac{L}{2}G_i^2$ from which we get $\frac{E_0}{\hbar\omega_c} = < -\frac{1}{2L}\left(-2\alpha + 4\alpha^2 G_i^2\right) + \frac{L}{2}G_i^2 >$. Using $a = 0$, $\sigma = 1.859 \times 10^{-6}$, $\beta = 0.8$, $k = 10^{-16}$, $L = 9.046 \times 10^{10}$, $\alpha = \beta L = 7.237 \times 10^{10}$, $\omega_c = 10.483$ MHz.

i	u_i	g_i	$G_i = a + \sigma g_i$	$-\frac{1}{2L}(-2\alpha + 4\alpha^2 G_i^2) + \frac{L}{2}G_i^2$
1	0.169	−0.955	−1.775E−06	5.776E−01
2	0.050	−1.635	−3.039E−06	1.482E−01
3	0.844	1.015	1.887E−06	5.488E−01
4	0.182	−0.905	−1.682E−06	6.003E−01
5	0.599	0.255	4.740E−07	7.841E−01
6	0.985	2.195	4.081E−06	−3.748E−01
7	0.726	0.605	1.125E−06	7.107E−01
8	0.148	−1.040	−1.933E−06	5.363E−01
9	0.788	0.805	1.496E−06	6.420E−01
10	0.698	0.520	9.667E−07	7.341E−01
11	0.010	−2.295	−4.266E−06	−4.843E−01
12	0.397	−0.255	−4.740E−07	7.841E−01
13	0.852	1.050	1.952E−06	5.312E−01
14	0.761	0.715	1.329E−06	6.753E−01
15	0.309	−0.495	−9.202E−07	7.403E−01
16	0.113	−1.205	−2.240E−06	4.459E−01
17	0.212	−0.795	−1.478E−06	6.459E−01
18	0.809	0.880	1.636E−06	6.112E−01
19	0.084	−1.370	−2.547E−06	3.423E−01
20	0.276	−0.590	−1.097E−06	7.151E−01
21	0.243	−0.695	−1.292E−06	6.822E−01
22	0.989	2.325	4.322E−06	−5.181E−01
23	0.508	0.020	3.718E−08	7.999E−01
24	0.817	0.910	1.692E−06	5.981E−01
25	0.174	−0.935	−1.738E−06	5.868E−01

i	u_i	g_i	$G_i = a + \sigma g_i$	$-\frac{1}{2L}(-2\alpha + 4\alpha^2 G_i{}^2) + \frac{L}{2}G_i{}^2$
26	0.225	−0.750	−1.394E−06	6.628E−01
27	0.687	0.490	9.109E−07	7.415E−01
28	0.060	−1.540	−2.863E−06	2.217E−01
29	0.070	−1.465	−2.723E−06	2.767E−01
30	0.835	0.980	1.822E−06	5.658E−01
31	0.466	−0.085	−1.580E−07	7.982E−01
32	0.271	−0.605	−1.125E−06	7.107E−01
33	0.813	0.895	1.664E−06	6.047E−01
34	0.376	−0.315	−5.856E−07	7.758E−01
35	0.983	2.145	3.988E−06	−3.219E−01
36	0.181	−0.905	−1.682E−06	6.003E−01
37	0.583	0.210	3.904E−07	7.892E−01
38	0.612	0.285	5.298E−07	7.802E−01
39	0.468	−0.080	−1.487E−07	7.984E−01
40	0.614	0.290	5.391E−07	7.795E−01
41	0.768	0.735	1.366E−06	6.683E−01
42	0.811	0.885	1.645E−06	6.090E−01
43	0.461	−0.095	−1.766E−07	7.978E−01
44	0.985	2.195	4.081E−06	−3.748E−01
45	0.735	0.630	1.171E−06	7.032E−01
46	0.559	0.150	2.789E−07	7.945E−01
47	0.884	1.200	2.231E−06	4.489E−01
48	0.229	−0.740	−1.376E−06	6.665E−01
49	0.920	1.415	2.630E−06	3.118E−01
50	0.524	0.060	1.115E−07	7.991E−01

i	u_i	g_i	$G_i = a + \sigma\, g_i$	$-\dfrac{1}{2L}(-2a + 4a^2 G_i^2) + \dfrac{L}{2} G_i^2$
51	0.393	−0.270	−5.019E−07	7.822E−01
52	0.306	−0.505	−9.388E−07	7.378E−01
53	0.819	0.915	1.701E−06	5.958E−01
54	0.620	0.310	5.763E−07	7.766E−01
55	0.875	1.155	2.147E−06	4.747E−01
56	0.773	0.750	1.394E−06	6.628E−01
57	0.492	−0.020	−3.718E−08	7.999E−01
58	0.332	−0.430	−7.994E−07	7.549E−01
59	0.185	−0.890	−1.655E−06	6.069E−01
60	0.565	0.165	3.067E−07	7.934E−01
61	0.434	−0.165	−3.067E−07	7.934E−01
62	0.665	0.430	7.994E−07	7.549E−01
63	0.516	0.040	7.436E−08	7.996E−01
64	0.659	0.410	7.622E−07	7.590E−01
65	0.709	0.555	1.032E−06	7.249E−01
66	0.515	0.040	7.436E−08	7.996E−01
67	0.683	0.480	8.923E−07	7.438E−01
68	0.092	−1.320	−2.454E−06	3.751E−01
69	0.417	−0.205	−3.811E−07	7.898E−01
70	0.598	0.250	4.648E−07	7.848E−01
71	0.682	0.475	8.830E−07	7.450E−01
72	0.948	1.635	3.039E−06	1.482E−01
73	0.125	−1.145	−2.129E−06	4.803E−01
74	0.149	−1.035	−1.924E−06	5.388E−01
75	0.246	−0.685	−1.273E−06	6.856E−01

i	u_i	g_i	$G_i = a + \sigma\, g_i$	$-\dfrac{1}{2L}(-2a + 4a^2 G_i^2) + \dfrac{L}{2} G_i^2$
76	0.231	−0.730	−1.357E−06	6.701E−01
77	0.836	0.985	1.831E−06	5.634E−01
78	0.167	−0.960	−1.785E−06	5.753E−01
79	0.545	0.115	2.138E−07	7.968E−01
80	0.748	0.670	1.246E−06	6.905E−01
81	0.440	−0.150	−2.789E−07	7.945E−01
82	0.674	0.455	8.458E−07	7.495E−01
83	0.287	−0.560	−1.041E−06	7.235E−01
84	0.052	−1.615	−3.002E−06	1.640E−01
85	0.940	1.565	2.909E−06	2.028E−01
86	0.679	0.470	8.737E−07	7.461E−01
87	0.910	1.350	2.510E−06	3.556E−01
88	0.254	−0.660	−1.227E−06	6.938E−01
89	0.236	−0.715	−1.329E−06	6.753E−01
90	0.854	1.060	1.971E−06	5.260E−01
91	0.877	1.165	2.166E−06	4.691E−01
92	0.483	−0.040	−7.436E−08	7.996E−01
93	0.117	−1.185	−2.203E−06	4.576E−01
94	0.847	1.030	1.915E−06	5.413E−01
95	0.201	−0.835	−1.552E−06	6.300E−01
96	0.294	−0.540	−1.004E−06	7.289E−01
97	0.071	−1.460	−2.714E−06	2.802E−01
98	0.065	−1.505	−2.798E−06	2.477E−01
99	0.740	0.645	1.199E−06	6.986E−01
100	0.954	1.695	3.151E−06	9.943E−02
				Sum = 57.058
				$E_0/(\hbar w_c) = 0.571$

Table 3.4: Showing values of ground state energy $E_0/(\hbar\omega_c)$ of an electron in simple harmonic oscillator potential $V(x) = \frac{1}{2}kx^2$ with $k = 10^{-16}$, for various values of variational parameter β contained in trial wavefunction $\Phi_0 = e^{-\beta L x^2} = e^{-\beta \frac{m\omega_c}{\hbar} x^2}$

β	$E_0/(\hbar w_c)$
0.1	1.229
0.2	0.694
0.3	0.551
0.4	0.506
0.5	0.500
0.6	0.514
0.8	0.571
1.0	0.647
1.2	0.733
1.4	0.825
1.6	0.921
1.8	1.019
2.0	1.117

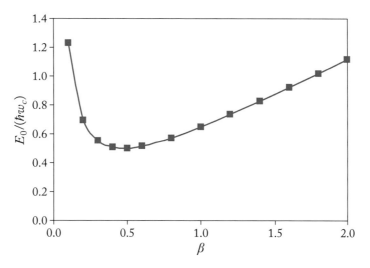

Figure 3.4: Showing values of ground state energy $E_0/(\hbar\omega_c)$ of an electron in simple harmonic oscillator potential $V(x) = \frac{1}{2}kx^2$ with $k = 10^{-16}$, for various values of variational parameter β contained in trial wavefunction $\Phi_0 = e^{-\beta L x^2} = e^{-\beta \frac{m\omega_c}{\hbar} x^2}$. Minimum value of $E_0/(\hbar\omega_c) = 0.5$ occurs for $\beta = 0.5$. Thus, the ground state eigenfunction is $\Phi_0 = e^{-\beta L x^2} = e^{-\frac{1}{2}Lx^2} = e^{-\frac{1}{2} \frac{m\omega_c}{\hbar} x^2}$ while ground state eigenvalue of energy is $E_0 = 0.5\hbar\omega_c$.

CHAPTER 4

Variational Monte Carlo Method Applied to the Ground State of a Hydrogen Atom

4.1 THE VARIATIONAL METHOD OF QUANTUM MECHANICS APPLIED TO THE GROUND STATE OF ANY QUANTUM MECHANICAL SYSTEM (AGAIN)

Let $\Psi_0, \Psi_1, \Psi_2, \Psi_3, \ldots, \Psi_N$ be exact, unknown, orthonormal eigenfunctions of known Hamiltonian H. Let $E_0 < E_1 < E_2 < E_3 < \ldots < E_N$ be exact, unknown eigenvalues of energy, respectively. That is, we have eigenvalue equation

$$H\Psi_n = E_n\Psi_n \tag{4.1}$$

which we cannot solve. Suppose we wish to estimate E_0 and we wish to know approximate form of Ψ_0.

Let us choose *by guess* a suitable function Φ_0 that is expected to resemble unknown Ψ_0 and write

$$\Phi_0 = \sum_{n=0}^{N} C_n^{(0)}\Psi_n \tag{4.2}$$

and let us evaluate the quantity E_{Φ_0} given by

$$E_{\Phi_0} = \frac{(\Phi_0, H\Phi_0)}{(\Phi_0, \Phi_0)} = \frac{\int \Phi_0^* H\Phi_0 \, d\tau}{\int \Phi_0^* \Phi_0 \, d\tau} \tag{4.3}$$

$$= \frac{\left(\sum_{m=0} C_m^{(0)}\Psi_m, H \sum_{n=0} C_n^{(0)}\Psi_n\right)}{\left(\sum_{m=0} C_m^{(0)}\Psi_m, \sum_{n=0} C_n^{(0)}\Psi_n\right)} = \frac{\left(\sum_{m=0} C_m^{(0)}\Psi_m, \sum_{n=0} C_n^{(0)} H\Psi_n\right)}{\left(\sum_{m=0} C_m^{(0)}\Psi_m, \sum_{n=0} C_n^{(0)}\Psi_n\right)}$$

$$= \frac{\left(\sum_{m=0} C_m^{(0)}\Psi_m, \sum_{n=0} C_n^{(0)} E_n\Psi_n\right)}{\left(\sum_{m=0} C_m^{(0)}\Psi_m, \sum_{n=0} C_n^{(0)}\Psi_n\right)}$$

using Equation (4.1)

$$= \frac{\sum_{m=0} \sum_{n=0} C_m^{(0)*} C_n^{(0)} E_n (\Psi_m, \Psi_n)}{\sum_{m=0} \sum_{n=0} C_m^{(0)*} C_n^{(0)} (\Psi_m, \Psi_n)} = \frac{\sum_{m=0} \sum_{n=0} C_m^{(0)*} C_n^{(0)} E_n \delta_{mn}}{\sum_{m=0} \sum_{n=0} C_m^{(0)*} C_n^{(0)} \delta_{mn}}$$

$$= \frac{\sum_{n=0} \left| C_n^{(0)} \right|^2 E_n}{\sum_{n=0} \left| C_n^{(0)} \right|^2}.$$

Thus,

$$E_{\Phi_0} - E_0 = \frac{\sum_{n=0} \left| C_n^{(0)} \right|^2 E_n}{\sum_{n=0} \left| C_n^{(0)} \right|^2} - E_0$$

$$= \frac{\sum_{n=0} \left| C_n^{(0)} \right|^2 (E_n - E_0)}{\sum_{n=0} \left| C_n^{(0)} \right|^2}.$$

(4.4)

In Equation (4.4), $E_n \geq E_0$. $\therefore E_n - E_0 \geq 0$; moreover, $\left| C_n^{(0)} \right|^2 \geq 0$. Thus, Equation (4.4) shows that $E_{\Phi_0} - E_0 \geq 0$ or, $E_{\Phi_0} \geq E_0$. We find that for arbitrary Φ_0, calculated value of $E_{\Phi_0} \geq E_0$, i.e., E_{Φ_0} is larger than or equal to ground state energy E_0. E_{Φ_0} gives upper bound of ground state energy E_0.

If chosen Φ_0 is different from actual unknown Ψ_0, calculated value of E_{Φ_0} calculated using Equation (4.3) will be larger than E_0. If chosen Φ_0 happens to be same as unknown Ψ_0, value of E_{Φ_0} calculated using Equation (4.3) becomes equal to E_0. Thus, we can choose arbitrary Φ_0 and calculate E_{Φ_0} using Equation (4.3) and get upper bound of E_0.

Usually we choose a suitable Φ_0 known as trail function containing one or more adjustable parameters $\alpha_i^{(0)}$'s, e.g., $\Phi_0 = e^{-\alpha_1^{(0)} x}$ where $\alpha_1^{(0)}$ is an adjustable parameter. Then we use this expression of Φ_0 in Equation (4.3) and calculated expression of E_{Φ_0} will contain the parameters $\alpha_i^{(0)}$'s. We then stipulate $\frac{\partial E_{\Phi_0}}{\partial \alpha_i^{(0)}} = 0$ for each $\alpha_i^{(0)}$ separately. We thus get values of $\alpha_i^{(0)}$'s for which E_{Φ_0} is minimum. We can put these values of $\alpha_i^{(0)}$'s in the expression of E_{Φ_0}. We then get the lowest value of E_{Φ_0} for the chosen Φ_0. This upper bound to E_0 will be close to actual unknown value of E_0 if trail function Φ_0 has a form closely resembling that of actual unknown Ψ_0. The entire process is repeated for several different functions as Φ_0 to see which function leads to much lower value of E_{Φ_0}. If we are happy that a calculated value of E_{Φ_0} is low enough and that it is likely to be close to actual unknown value of E_0, we can take $E_0 \approx E_{\Phi_0}$ and $\Psi_0 \approx \Phi_0$ within some approximation.

4.2   THE VARIATIONAL METHOD OF QUANTUM MECHANICS APPLIED TO THE GROUND STATE OF A HYDROGEN ATOM

We have a Hamiltonian operator $H = -\frac{\hbar^2}{2m}\nabla^2 - \frac{e^2}{4\pi\varepsilon_0}\frac{1}{r}$ of a hydrogen atom. More explicitly, we have in a spherical polar coordinate system the Hamiltonian operator as

$$H = -\frac{\hbar^2}{2m}\left[\frac{1}{r^2}\frac{\partial}{\partial r}\left(r^2\frac{\partial}{\partial r}\right) + \frac{1}{r^2\mathrm{Sin}\theta}\frac{\partial}{\partial\theta}\left(\mathrm{Sin}\theta\frac{\partial}{\partial\theta}\right) + \frac{1}{r^2\mathrm{Sin}^2\theta}\frac{\partial^2}{\partial\phi^2}\right] - \frac{e^2}{4\pi\varepsilon_0}\frac{1}{r},$$

where $m = 9.095 \times 10^{-31}$ kg is a reduced mass of a proton-electron two-body system.

Let us choose $\Phi_0 = e^{-\alpha r} = e^{-r/a_0}$ by guess as an approximate form of ground state eigenfunction Φ_0 and let us evaluate the quantity

$$E_0 = \frac{(\Phi_0, H\Phi_0)}{(\Phi_0, \Phi_0)}. \tag{4.5}$$

Here, $a_0 = 0.53$ Å is the so-called Bohr radius. Now,

$$(\Phi_0, \Phi_0) = (e^{-\alpha r}, e^{-\alpha r})$$

$$= \int_{\phi=0}^{2\pi}\int_{\theta=0}^{\pi}\int_{r=0}^{\infty} e^{-2\alpha r}r^2 dr\,\mathrm{Sin}\theta d\theta d\phi = 4\pi\int_{r=0}^{\infty} e^{-2\alpha r}r^2 dr$$

$$= 4\pi\int_{r=0}^{\infty} e^{-(2\alpha)r}r^{3-1}dr = 4\pi(2\alpha)^{-3}2! = \frac{\pi}{\alpha^3}.$$

$$(\Phi_0, \Phi_0) = \frac{\pi}{\alpha^3}. \tag{4.6}$$

Now,

$$H\Phi_0 = -\frac{\hbar^2}{2m}\left[\frac{1}{r^2}\frac{\partial}{\partial r}\left(r^2\frac{\partial\Phi_0}{\partial r}\right) + \frac{1}{r^2\mathrm{Sin}\theta}\frac{\partial}{\partial\theta}\left(\mathrm{Sin}\theta\frac{\partial\Phi_0}{\partial\theta}\right) + \frac{1}{r^2\mathrm{Sin}^2\theta}\frac{\partial^2\Phi_0}{\partial\phi^2}\right] - \frac{e^2}{4\pi\varepsilon_0}\frac{1}{r}\Phi_0$$

or

$$H\Phi_0 = -\frac{\hbar^2}{2m}\left[\frac{1}{r^2}\frac{\partial}{\partial r}\left(r^2\frac{\partial\Phi_0}{\partial r}\right) + 0 + 0\right] - \frac{e^2}{4\pi\varepsilon_0}\frac{1}{r}\Phi_0$$

or

$$H\Phi_0 = -\frac{\hbar^2}{2m}\left[\frac{1}{r^2}\frac{\partial}{\partial r}\left(r^2\frac{\partial\Phi_0}{\partial r}\right)\right] - \frac{e^2}{4\pi\varepsilon_0}\frac{1}{r}\Phi_0$$

or

$$H\Phi_0 = -\frac{\hbar^2}{2m}\left[\frac{1}{r^2}\frac{\partial}{\partial r}\left(r^2\frac{\partial e^{-\alpha r}}{\partial r}\right)\right] - \frac{e^2}{4\pi\varepsilon_0}\frac{1}{r}e^{-\alpha r}$$

or

$$H\Phi_0 = -\frac{\hbar^2}{2m}\left[\alpha^2 - \frac{2}{r}\alpha\right]e^{-\alpha r} - \frac{e^2}{4\pi\varepsilon_0}\frac{1}{r}e^{-\alpha r}. \tag{4.7}$$

As such,

$$E_0 = \frac{(\Phi_0, H\Phi_0)}{(\Phi_0, \Phi_0)}$$

gives

$$E_0 = \frac{(\Phi_0, H\Phi_0)}{\pi/\alpha^3},$$

or

$$E_0 = \frac{4\pi \int_{r=0}^{\infty}\left(-\frac{\hbar^2}{2m}\left[\alpha^2 - \frac{2}{r}\alpha\right] - \frac{e^2}{4\pi\varepsilon_0}\frac{1}{r}\right)e^{-2\alpha r}r^2 dr}{\pi/\alpha^3}$$

or

$$E_0 = \int_{r=0}^{\infty} 4\alpha^3 \left(-\frac{\hbar^2}{2m}\left[\alpha^2 - \frac{2}{r}\alpha\right]e^{-2\alpha r} - \frac{e^2}{4\pi\varepsilon_0}\frac{1}{r}e^{-2\alpha r}\right)r^2 dr. \tag{4.8}$$

This integral is evaluated in the next section using the Monte Carlo method.

4.3   GROUND STATE ENERGY OF A HYDROGEN ATOM USING THE VARIATIONAL MONTE CARLO METHOD

We first gather the *exact* results from the Quantum Mechanics books. The ground state eigen-function is $\Phi_0 = e^{-\alpha r} = e^{-r/a_0}$ with a_0 called Bohr radius given by 0.53 Å and eigenvalue of energy equal to -13.6 eV.

As to the Monte Carlo method, we take up Equation (4.8):

$$E_0 = \int_{r=0}^{\infty} 4\alpha^3 \left(-\frac{\hbar^2}{2m}\left[\alpha^2 - \frac{2}{r}\alpha\right]e^{-2\alpha r} - \frac{e^2}{4\pi\varepsilon_0}\frac{1}{r}e^{-2\alpha r}\right)r^2 dr,$$

and rewrite it as

$$E_0 = \int_{r=0}^{\infty} F(r)dr, \tag{4.9}$$

where

$$F(r) = 4\alpha^3 \left(-\frac{\hbar^2}{2m}\left[\alpha^2 - \frac{2}{r}\alpha\right]e^{-2\alpha r} - \frac{e^2}{4\pi\varepsilon_0}\frac{1}{r}e^{-2\alpha r}\right)r^2. \tag{4.10}$$

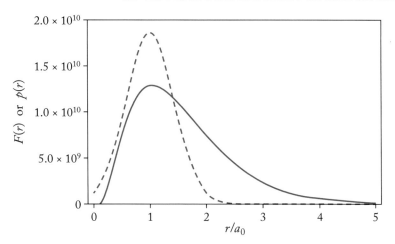

Figure 4.1: Smooth curve shows $-0.55 \times 10^{18} F(r)$ and dashed curve shows Gaussian $p(r) = \frac{1}{\sigma \sqrt{2\pi}} \mathrm{Exp}\left[-\frac{1}{2}\left(\frac{r-a}{\sigma}\right)^2\right]$ for $a = 1.0a_0$ and $\sigma = 0.43a_0, a_0 = 0.5$ Å. We find that $F(r)$ and Gaussian $p(r)$ are nearly proportional to each other and hence Gaussian $p(r)$ is an ideal choice for the Monte Carlo calculation.

We now re-write Equation (4.9) as

$$E_0 = \int_{r=0}^{\infty} \frac{F(r)}{p(r)} p(r) dr, \tag{4.11}$$

where $p(r)$ is a probability density function to be taken such that it resembles $F(r)$. Thus, E_0 is given by average or expectation value of $\frac{F(r)}{p(r)}$, i.e.,

$$E_0 = < \frac{F(r)}{p(r)} > . \tag{4.12}$$

Thus, what we need to do is to see the functional form of $F(r)$ and get a probability density function $p(r)$ that looks like $F(r)$ and calculate average value of $\frac{F(r)}{p(r)}$ for various values of a_0 in the range 0.3–1 Å. We need to find out the lowest value of $< \frac{F(r)}{p(r)} >$ which is expected to yield about -13.6 eV.

Using Program 4.1 in Mathematica 6.0, for $a_0 = 0.5$ Å, we have plotted $-0.55 \times 10^{18} F(r)$ and Gaussian

$$p(r) = \frac{1}{\sigma \sqrt{2\pi}} \mathrm{Exp}\left[-\frac{1}{2}\left(\frac{r-a}{\sigma}\right)^2\right]$$

for $a = 1.0a_0$ and $\sigma = 0.43a_0$. We find that $F(r)$ and Gaussian $p(r)$ are nearly proportional to each other and hence Gaussian $p(r)$ is an ideal choice for the Monte Carlo calculation.

Program 4.1

```
a0=0.5*10^-10;
alpha=1/a0;
m=1837*(9.1*(10^-31))/1838;
hcut=(6.626*(10^-34))/(2*3.1416);
ce=1.6*(10^-19);
ep0=8.85*(10^-12);

F=Plot[(-0.55*10^18)*(4*(alpha^3)*(-(hcut^2)/
(2*m))*(alpha^2-2*alpha/(r*a0))*
Exp[-2*alpha*(r*a0)]-(4*(alpha^3))*((ce^2)/(4*3.1416*ep0*(r*a0)))*
Exp[-2*alpha*(r*a0)])*(r*a0)^2,{r,0,5},PlotStyle->{Black},Frame->True,
FrameLabel->{"r/a0","F(r) or p(r)"},PlotRange->{0,2*10^10}]

sig=0.43*a0;
a=1.0*a0;
p=Plot[(1/(sig*Sqrt[2*Pi]))*Exp[-0.5*((r*a0-a)/sig)^2],{r,0,5},
PlotStyle->{Dashed,Black},Frame->True,
FrameLabel->{"r/a0","F(r) or p(r)"}, PlotRange->{0,2*10^10}]
Show[F,p]
```

For values of a_0 in the range 0.3–1 Å, we have run the Program 4.1 and find that $F(r)$ and $p(r)$ match well for different pairs of values of a and σ tabulated in Table 4.1.

Using Table 4.2 in Microsoft Excel, we have evaluated the expectation value set forth in Equation (4.12): $E_0 =< \frac{F(r)}{p(r)} >$ as a value of the ground state energy. Here,

$$F(r) = 4\alpha^3 \left(-\frac{\hbar^2}{2m} \left[\alpha^2 - \frac{2}{r}\alpha \right] e^{-2\alpha r} - \frac{e^2}{4\pi\varepsilon_0}\frac{1}{r}e^{-2\alpha r} \right) r^2$$

and

$$p(r) = \frac{1}{\sigma\sqrt{2\pi}} \text{Exp}\left[-\frac{1}{2}\left(\frac{r-a}{\sigma}\right)^2 \right].$$

We have run Program 4.1 and executed the Microsoft Excel spreadsheet for Table 4.2 for various values of a_0 namely 0.3–1.0 Å and obtained values of E_0 in each case. Table 4.3 shows the results which are plotted in Figure 4.2. We find from Figure 4.2 that E_0 reaches a minimum for a_0 around 0.5 Å as expected. The minimum value is about −11.34 eV for $a_0 =$

Table 4.1: Pairs of values of a and σ for which $F(r)$ and $p(r)$ match with each other within a constant multiple, ascertained using Program 4.1

a_0 in Å	a	σ
0.3	$1.6a_0$	$0.50a_0$
0.4	$1.3a_0$	$0.50a_0$
0.5	$1.0a_0$	$0.43a_0$
0.6	$0.9a_0$	$0.40a_0$
0.7	$0.8a_0$	$0.35a_0$
0.8	$0.7a_0$	$0.35a_0$
0.9	$0.7a_0$	$0.30a_0$
1.0	$0.6a_0$	$0.25a_0$

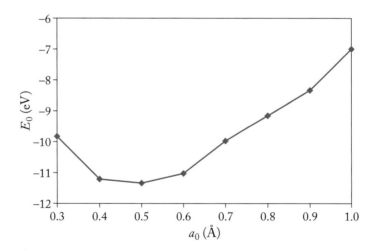

Figure 4.2: Showing values of ground state energy E_0 of a hydrogen atom for various values of Bohr radius a_0 contained in trial wavefunction $\Phi_0 = e^{-r/a_0}$. Minimum value of $E_0 = -11.34$ eV occurs around $a_0 = 0.5$ Å.

0.5 Å. The deviation from -13.6 eV is not surprising given only 100 random numbers used in the calculations.

Table 4.2: Showing tabulated values of u_i, g_i, G_i, $F(G_i) = 4\alpha^3 \left(-\frac{\hbar^2}{2m} \left[\alpha^2 - \frac{2}{G_i}\alpha \right] e^{-2\alpha G_i} \right.$ $\left. -\frac{e^2}{4\pi\varepsilon_0} \frac{1}{G_i} e^{-2\alpha G_i} \right) G_i^2$, and $p(G_i) = \frac{1}{\sigma\sqrt{2\pi}} \mathrm{Exp} \left[-\frac{1}{2} \left(\frac{G_i - a}{\sigma} \right)^2 \right]$, $F(G_i)/p(G_i)$ from which we get the average value of $E_0 = < \frac{F}{p} >$ as a value of the ground state energy. Using $a_0 = 0.5$ Å, $a = 1.0a_0$, $\sigma = 0.43a_0$.

i	u_i	g_i	G_i	F_i	p_i	F_i/p_i
1	0.169	−0.955	2.95E−11	−1.67E−08	1.18E+10	−1.42E−18
2	0.050	−1.635	1.48E−11	−5.76E−09	4.88E+09	−1.18E−18
3	0.844	1.015	7.18E−11	−2.10E−08	1.11E+10	−1.89E−18
4	0.182	−0.905	3.05E−11	−1.74E−08	1.23E+10	−1.41E−18
5	0.599	0.255	5.55E−11	−2.34E−08	1.80E+10	−1.30E−18
6	0.985	2.195	9.72E−11	−1.42E−08	1.67E+09	−8.53E−18
7	0.726	0.605	6.30E−11	−2.27E−08	1.55E+10	−1.47E−18
8	0.148	−1.040	2.76E−11	−1.56E−08	1.08E+10	−1.44E−18
9	0.788	0.805	6.73E−11	−2.19E−08	1.34E+10	−1.63E−18
10	0.698	0.520	6.12E−11	−2.29E−08	1.62E+10	−1.41E−18
11	0.010	−2.295	6.58E−13	2.61E−10	1.33E+09	1.96E−19
12	0.397	−0.255	4.45E−11	−2.27E−08	1.80E+10	−1.26E−18
13	0.852	1.050	7.26E−11	−2.08E−08	1.07E+10	−1.94E−18
14	0.761	0.715	6.54E−11	−2.23E−08	1.44E+10	−1.55E−18
15	0.309	−0.495	3.94E−11	−2.14E−08	1.64E+10	−1.30E−18
16	0.113	−1.205	2.41E−11	−1.31E−08	8.98E+09	−1.46E−18
17	0.212	−0.795	3.29E−11	−1.87E−08	1.35E+10	−1.38E−18
18	0.809	0.880	6.89E−11	−2.16E−08	1.26E+10	−1.71E−18
19	0.084	−1.370	2.05E−11	−1.04E−08	7.26E+09	−1.43E−18
20	0.276	−0.590	3.73E−11	−2.06E−08	1.56E+10	−1.32E−18
21	0.243	−0.695	3.51E−11	−1.97E−08	1.46E+10	−1.35E−18
22	0.989	2.325	1.00E−10	−1.35E−08	1.24E+09	−1.09E−17
23	0.508	0.020	5.04E−11	−2.34E−08	1.86E+10	−1.26E−18
24	0.817	0.910	6.96E−11	−2.15E−08	1.23E+10	−1.75E−18
25	0.174	−0.935	2.99E−11	−1.70E−08	1.20E+10	−1.42E−18

i	u_i	g_i	G_i	F_i	p_i	F_i/p_i
26	0.225	−0.750	3.39E−11	−1.91E−08	1.40E+10	−1.37E−18
27	0.687	0.490	6.05E−11	−2.30E−08	1.65E+10	−1.40E−18
28	0.060	−1.540	1.69E−11	−7.41E−09	5.67E+09	−1.31E−18
29	0.070	−1.465	1.85E−11	−8.72E−09	6.34E+09	−1.38E−18
30	0.835	0.980	7.11E−11	−2.11E−08	1.15E+10	−1.84E−18
31	0.466	−0.085	4.82E−11	−2.32E−08	1.85E+10	−1.26E−18
32	0.271	−0.605	3.70E−11	−2.05E−08	1.55E+10	−1.33E−18
33	0.813	0.895	6.92E−11	−2.15E−08	1.24E+10	−1.73E−18
34	0.376	−0.315	4.32E−11	−2.24E−08	1.77E+10	−1.27E−18
35	0.983	2.145	9.61E−11	−1.45E−08	1.86E+09	−7.81E−18
36	0.181	−0.905	3.05E−11	−1.74E−08	1.23E+10	−1.41E−18
37	0.583	0.210	5.45E−11	−2.34E−08	1.82E+10	−1.29E−18
38	0.612	0.285	5.61E−11	−2.34E−08	1.78E+10	−1.31E−18
39	0.468	−0.080	4.83E−11	−2.32E−08	1.85E+10	−1.26E−18
40	0.614	0.290	5.62E−11	−2.34E−08	1.78E+10	−1.31E−18
41	0.768	0.735	6.58E−11	−2.22E−08	1.42E+10	−1.57E−18
42	0.811	0.885	6.90E−11	−2.16E−08	1.25E+10	−1.72E−18
43	0.461	−0.095	4.80E−11	−2.32E−08	1.85E+10	−1.26E−18
44	0.985	2.195	9.72E−11	−1.42E−08	1.67E+09	−8.53E−18
45	0.735	0.630	6.35E−11	−2.26E−08	1.52E+10	−1.48E−18
46	0.559	0.150	5.32E−11	−2.35E−08	1.83E+10	−1.28E−18
47	0.884	1.200	7.58E−11	−2.00E−08	9.03E+09	−2.21E−18
48	0.229	−0.740	3.41E−11	−1.93E−08	1.41E+10	−1.36E−18
49	0.920	1.415	8.04E−11	−1.88E−08	6.82E+09	−2.76E−18
50	0.524	0.060	5.13E−11	−2.34E−08	1.85E+10	−1.27E−18

i	u_i	g_i	G_i	F_i	p_i	F_i/p_i
51	0.393	−0.270	4.42E−11	−2.26E−08	1.79E+10	−1.26E−18
52	0.306	−0.505	3.91E−11	−2.13E−08	1.63E+10	−1.30E−18
53	0.819	0.915	6.97E−11	−2.14E−08	1.22E+10	−1.76E−18
54	0.620	0.310	5.67E−11	−2.33E−08	1.77E+10	−1.32E−18
55	0.875	1.155	7.48E−11	−2.02E−08	9.52E+09	−2.13E−18
56	0.773	0.750	6.61E−11	−2.21E−08	1.40E+10	−1.58E−18
57	0.492	−0.020	4.96E−11	−2.33E−08	1.86E+10	−1.26E−18
58	0.332	−0.430	4.08E−11	−2.18E−08	1.69E+10	−1.29E−18
59	0.185	−0.890	3.09E−11	−1.76E−08	1.25E+10	−1.41E−18
60	0.565	0.165	5.35E−11	−2.35E−08	1.83E+10	−1.28E−18
61	0.434	−0.165	4.65E−11	−2.30E−08	1.83E+10	−1.26E−18
62	0.665	0.430	5.92E−11	−2.31E−08	1.69E+10	−1.37E−18
63	0.516	0.040	5.09E−11	−2.34E−08	1.85E+10	−1.26E−18
64	0.659	0.410	5.88E−11	−2.32E−08	1.71E+10	−1.36E−18
65	0.709	0.555	6.19E−11	−2.28E−08	1.59E+10	−1.43E−18
66	0.515	0.040	5.09E−11	−2.34E−08	1.85E+10	−1.26E−18
67	0.683	0.480	6.03E−11	−2.30E−08	1.65E+10	−1.39E−18
68	0.092	−1.320	2.16E−11	−1.12E−08	7.76E+09	−1.45E−18
69	0.417	−0.205	4.56E−11	−2.29E−08	1.82E+10	−1.26E−18
70	0.598	0.250	5.54E−11	−2.34E−08	1.80E+10	−1.30E−18
71	0.682	0.475	6.02E−11	−2.30E−08	1.66E+10	−1.39E−18
72	0.948	1.635	8.52E−11	−1.75E−08	4.88E+09	−3.59E−18
73	0.125	−1.145	2.54E−11	−1.40E−08	9.63E+09	−1.46E−18
74	0.149	−1.035	2.77E−11	−1.57E−08	1.09E+10	−1.44E−18
75	0.246	−0.685	3.53E−11	−1.98E−08	1.47E+10	−1.35E−18

i	u_i	g_i	G_i	F_i	p_i	F_i/p_i
76	0.231	−0.730	3.43E−11	−1.94E−08	1.42E+10	−1.36E−18
77	0.836	0.985	7.12E−11	−2.11E−08	1.14E+10	−1.85E−18
78	0.167	−0.960	2.94E−11	−1.67E−08	1.17E+10	−1.42E−18
79	0.545	0.115	5.25E−11	−2.35E−08	1.84E+10	−1.27E−18
80	0.748	0.670	6.44E−11	−2.24E−08	1.48E+10	−1.51E−18
81	0.440	−0.150	4.68E−11	−2.31E−08	1.83E+10	−1.26E−18
82	0.674	0.455	5.98E−11	−2.31E−08	1.67E+10	−1.38E−18
83	0.287	−0.560	3.80E−11	−2.09E−08	1.59E+10	−1.32E−18
84	0.052	−1.615	1.53E−11	−6.10E−09	5.04E+09	−1.21E−18
85	0.940	1.565	8.36E−11	−1.79E−08	5.45E+09	−3.29E−18
86	0.679	0.470	6.01E−11	−2.30E−08	1.66E+10	−1.39E−18
87	0.910	1.350	7.90E−11	−1.92E−08	7.46E+09	−2.57E−18
88	0.254	−0.660	3.58E−11	−2.00E−08	1.49E+10	−1.34E−18
89	0.236	−0.715	3.46E−11	−1.95E−08	1.44E+10	−1.36E−18
90	0.854	1.060	7.28E−11	−2.07E−08	1.06E+10	−1.96E−18
91	0.877	1.165	7.50E−11	−2.02E−08	9.41E+09	−2.14E−18
92	0.483	−0.040	4.91E−11	−2.33E−08	1.85E+10	−1.26E−18
93	0.117	−1.185	2.45E−11	−1.34E−08	9.19E+09	−1.46E−18
94	0.847	1.030	7.21E−11	−2.09E−08	1.09E+10	−1.91E−18
95	0.201	−0.835	3.20E−11	−1.82E−08	1.31E+10	−1.39E−18
96	0.294	−0.540	3.84E−11	−2.10E−08	1.60E+10	−1.31E−18
97	0.071	−1.460	1.86E−11	−8.81E−09	6.39E+09	−1.38E−18
98	0.065	−1.505	1.76E−11	−8.02E−09	5.98E+09	−1.34E−18
99	0.740	0.645	6.39E−11	−2.25E−08	1.51E+10	−1.49E−18
100	0.954	1.695	8.64E−11	−1.72E−08	4.41E+09	−3.89E−18
					Sum = −1.81E−16	
					E_0 = −11.33 eV	

Table 4.3: Showing values of ground state energy E_0 of a hydrogen atom for various values of Bohr radius a_0 contained in trial wavefunction $\Phi_0 = e^{-r/a_0}$

a_0 (Å)	E_0 (eV)
0.3	−9.83
0.4	−11.21
0.5	**−11.34**
0.6	−11.03
0.7	−9.98
0.8	−9.16
0.9	−8.34
1.0	−6.98

CHAPTER 5

Concluding Remarks

1. This book is intended for undergraduate students of Mathematics, Statistics, and Physics who know nothing about the Monte Carlo method but wish to know how it works.

2. All treatments have been done as much manually as is practicable.

3. The treatments are deliberately manual to let the readers get the real feel of how the Monte Carlo method works.

4. Definite integrals of a total of five functions $F(x)$, namely $\mathrm{Sin}(x), \mathrm{Cos}(x), e^x, \log_e(x)$, $\frac{1}{1+x^2}$, have been evaluated using constant, linear, Gaussian, and exponential probability density functions $p(x)$.

5. It is shown that results agree with known exact values better if $p(x)$ is proportional to $F(x)$.

6. Deviation from the proportionality results in worse agreement.

7. Necessary background materials are covered in Chapter 1 while the integrals are evaluated in Chapter 2.

8. Two separate chapters have been dedicated to the variational Monte Carlo method applied to the ground state of a simple harmonic oscillator and of a hydrogen atom with remarkable success given only 100 random numbers used.

9. The book is a good read, and is intended to make the readers adept at using the method.

Bibliography

[1] IIya M. Sobol, *A Primer for the Monte Carlo Method*, CRC Press, 1994.

[2] IIya M. Sobol, *The Monte Carlo Method*, The University of Chicago Press, 1974.

[3] Yu. A. Shreider, Ed., *The Monte Carlo Method: The Method of Statistical Trials*, Pergamon Press, 1966.

Author's Biography

SUJAUL CHOWDHURY

Sujaul Chowdhury is a Professor in the Department of Physics at the Shahjalal University of Science and Technology (SUST), Sylhet, Bangladesh (www.sust.edu). He obtained a B.Sc. (Honours) in Physics in 1994 and an M.Sc. in Physics in 1996 from SUST. He obtained a Ph.D. in Physics from The University of Glasgow, UK in 2001. He was a Humboldt Research Fellow for one year at The Max Planck Institute, Stuttgart, Germany.

Printed in the United States
by Baker & Taylor Publisher Services

Scalability Challenges in
Web Search Engines

Synthesis Lectures on Information Concept, Retrieval, and Services

Editor
Gary Marchionini, *University of North Carolina, Chapel Hill*

Synthesis Lectures on Information Concepts, Retrieval, and Services publishes short books on topics pertaining to information science and applications of technology to information discovery, production, distribution, and management. Potential topics include: data models, indexing theory and algorithms, classification, information architecture, information economics, privacy and identity, scholarly communication, bibliometrics and webometrics, personal information management, human information behavior, digital libraries, archives and preservation, cultural informatics, information retrieval evaluation, data fusion, relevance feedback, recommendation systems, question answering, natural language processing for retrieval, text summarization, multimedia retrieval, multilingual retrieval, and exploratory search.

Scalability Challenges in Web Search Engines
B. Barla Cambazoglu and Ricardo Baeza-Yates
2015

Social Informatics Evolving
Pnina Fichman, Madelyn R. Sanfilippo, and Howard Rosenbaum
2015

On the Efficient Determination of Most Near Neighbors: Horseshoes, Hand Grenades, Web Search and Other Situations When Close Is Close Enough, Second Edition
Mark S. Manasse
2015

Building a Better World with Our Information: The Future of Personal Information Management, Part 3
William Jones
2015

Click Models for Web Search
Aleksandr Chuklin, Ilya Markov, and Maarten de Rijke
2015

Scalability Challenges in Web Search Engines
B. Barla Cambazoglu and Ricardo Baeza-Yates

ISBN: 978-3-031-01170-2   paperback
ISBN: 978-3-031-02298-2   ebook

DOI 10.1007/978-3-031-02298-2

A Publication in the Springer series
SYNTHESIS LECTURES ON INFORMATION CONCEPT, RETRIEVAL, AND SERVICES

Lecture #45
Series Editor: Gary Marchionini, *University of North Carolina, Chapel Hill*
Series ISSN
Print 1947-945X   Electronic 1947-9468

Scalability Challenges in
Web Search Engines

B. Barla Cambazoglu

Ricardo Baeza-Yates

SYNTHESIS LECTURES ON INFORMATION CONCEPT, RETRIEVAL, AND SERVICES #45

ABSTRACT

In this book, we aim to provide a fairly comprehensive overview of the scalability and efficiency challenges in large-scale web search engines. More specifically, we cover the issues involved in the design of three separate systems that are commonly available in every web-scale search engine: web crawling, indexing, and query processing systems. We present the performance challenges encountered in these systems and review a wide range of design alternatives employed as solution to these challenges, specifically focusing on algorithmic and architectural optimizations. We discuss the available optimizations at different computational granularities, ranging from a single computer node to a collection of data centers. We provide some hints to both the practitioners and theoreticians involved in the field about the way large-scale web search engines operate and the adopted design choices. Moreover, we survey the efficiency literature, providing pointers to a large number of relatively important research papers. Finally, we discuss some open research problems in the context of search engine efficiency.

KEYWORDS

cache invalidation, central broker, compression, content spam, delay attacks, distributed crawling, distributed query processing, DNS cache, document id reassignment, download throughput, dynamic index pruning, early exit optimization, effectiveness, efficiency, forward index, index construction, index maintenance, index partitioning, index replication, indexing, inverted index, inverted list cache, inverted list, link exchange, link farm, link spam, machine-learned ranking, matching, multisite web search, near duplicate detection, page cache, performance, position list, posting list, query-independent feature, query expansion, query forwarding, query interpretation, query processing, query rewriting, query scheduling, relevance, response latency, result cache, result freshness, result preparation, result retrieval, scalability, search center, search cluster, search engine result page, search quality, selective search, shingles, skip pointer, snippet, soft 404 page, spider trap, static index pruning, text processing, throughput, tiering, time-to-live, two-phase ranking, URL-seen test, URL caching, web change, web coverage, web crawler, web frontier, web graph, web repository, web search engine, website mirror

In loving memory of our grandmothers
Mübeccel and Nina.

Contents

Preface

This book aims to provide a comprehensive survey of the efficiency and scalability challenges in large-scale web search engines and the solutions adopted as a remedy. Like most other technical books, reading this book requires a certain level of prior knowledge, and its benefits will vary from reader to reader. To provide some guidance on who should read this book, we make the following recommendations (according to the readers' level of knowledge in web search):

Basic (e.g., an undergraduate student with basic knowledge of web search engines): The book provides limited preliminary information as it assumes its readers to have a good understanding of basic web search concepts. For readers who have just started to expand their knowledge in the field, an introductory information retrieval book may be a better option. Nevertheless, the book should be an easy read even for readers with basic knowledge as it avoids deep technical discussions as much as possible.

Intermediate (e.g., a post-graduate student in information retrieval or related areas): This type of readers are most likely to benefit from the book. We believe that the content of the book will let these readers expand their existing knowledge while potentially helping them to complete some missing links in their heads. Moreover, the extensive list of references provided in the book may help these readers to discover new information sources about specific topics.

Advanced (e.g., an information retrieval practitioner): As we mentioned, the book does not aim to dive into too much technical detail. However, it provides a comprehensive coverage of the field, touching a large number of practical issues in web search engine efficiency. Therefore, we hope that even readers with advanced knowledge will find something useful or interesting in the book.

We would like to thank Morgan & Claypool Publishers for providing us with the opportunity to write this book. In particular, we thank Gary Marchionini and Diane Cerra for their guidance and pushing this work to completion.

B. Barla Cambazoglu and Ricardo Baeza-Yates
November 2015

CHAPTER 1

Introduction

With the rapid growth of the Web and its wide-spread use in the last two decades, commercial web search engines have become indispensable tools in our lives. Today, we rely on search engines to satisfy almost every information need we have, from collecting references for a scientific paper to deciding what to cook for dinner. With the quick adoption of mobile technologies, search has become even more ubiquitous. Now, we can search on our phones or on our tablets, no matter where we are and what we are doing. In short, web search has become a key technology for quick access to information, and it is reasonable to expect that it will continue to matter in the foreseeable future (perhaps together with the communication technologies).

People use a search engine for many purposes. They do not only search for information, but also for other needs, such as finding the address of a website for easy navigation or purchasing a product online. In fact, today, most web search queries are navigational in nature, and those with an explicit information need form a small portion of the query volume. Nevertheless, it is safe to assume that all web search queries carry some sort of informational need, no matter if the need is explicit or implicit. Therefore, in the rest of the book, we will use the phrase "information need" in this general sense, which is broader than what information need would mean to a librarian.

The number of widely used web search engines we have today is less than a handful. This is mainly due to the fact that operating a large-scale web search engine is an expensive business. In practice, a web search engine needs to make continuous investment in hardware to cope with the growth of the Web and deal with the ever-increasing search volumes. But, investing in hardware alone is not a scalable solution in the long term. Designing efficient and scalable functional components is crucial to reduce the operational costs. In this book, we focus on the techniques and optimizations that are devised to improve the efficiency and scalability of web search engines.

Given the popularity of commercial search engines, one may be surprised to see how little is actually revealed to the general public about the machinery behind the web search business. Although a limited amount of behind-the-scenes knowledge is made available by the employees of search engine companies, occasionally through the posts on web blogs or public talks [Dean, 2009], most knowledge remains in the form of highly guarded secrets that are locked behind the gates of commercial search engine companies. We believe that this is mainly because the highly monetizable nature of the web search business and the tough competition in the search market render the techniques and engineering practices employed in search engines highly confidential.

One of the motivations when we started writing this book was to provide a complete picture of the scalability challenges in large-scale web search engines. In the mean time, we surveyed the publicly available information in the context of web search, aiming to provide an up-to-date coverage of the existing research literature. As one can imagine, we do not have endless freedom in what information we can reveal and to what extent, due to the confidentiality concerns mentioned above. Nevertheless, we believe that the information provided in this book is a current and accurate description of what really happens in practice.

The content of our book is based on a book chapter we published earlier [Cambazoglu and Baeza-Yates, 2011] as well as a series of tutorials we presented at top-tier conferences on information retrieval and web search, including SIGIR'13 [Cambazoglu and Baeza-Yates, 2013], WWW'14 [Baeza-Yates and Cambazoglu, 2014], SIGIR'14 [Cambazoglu and Baeza-Yates, 2014], WSDM'15 [Cambazoglu and Baeza-Yates, 2015], and CIKM'15. The structure of the book is quite simple. In this chapter, we first begin by giving a somewhat high-level overview of the concepts involved in web search. We then gradually restrict our scope to the main theme of the book: the scalability and efficiency challenges in web search engines. The chapter finishes with a literature survey on web search. The main chapters of the book make a deeper dive into the three functional systems that are present in every web search engine, namely, the web crawling system (Chapter 2), indexing system (Chapter 3), and query processing system (Chapter 4). The book ends in Chapter 5 with some concluding remarks and pointers to further reading.

1.1 WEB SEARCH BUSINESS

Perhaps the quickest way to get a good understanding of the web search business is to start by reviewing the constituents of a search engine result page (SERP). SERPs are the user-facing side of search engines and can let us visualize certain backend system tasks better. In Figure 1.1, we display three SERPs, each generated to answer the same search query ("buy darth vader mask"), but by different commercial web search engines. As we can see, all three SERPs bear some resemblance to each other. For example, all three SERPs provide a text box where users type their search queries as keywords. This box provides the basic means for users to communicate their information needs to the search engine. After the search query is entered, it is communicated to the search engine, where the query first goes through an interpretation process and then the web pages matching the identified information need are retrieved. The results are returned to the user in the form of a ranked list of web pages, sorted in decreasing order of pages' estimated usefulness or relevance to the identified information need. Generating these so-called algorithmic search results or "ten blue links" is at the heart of web search, and the degree to which these results sat-

(a) Google search results.

(b) Bing search results.

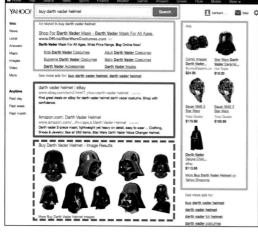

(c) Yahoo search results.

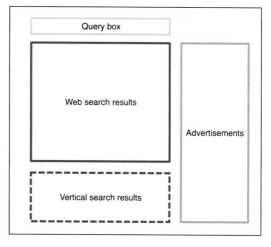

(d) Mapping between the boxes and modules.

Figure 1.1: Example modules in search engine result pages generated by Google, Bing, and Yahoo (the three search engines that have the largest traffic share in the U.S. search market as of November 2015).

isfy the user's information need is the main criterion when judging the service quality or, more generally, the success of a commercial web search engine.

Besides the web search result module, which is present in most SERPs (some queries do not match any results, in which case the web search result module is empty), there are also some auxiliary result modules that are occasionally presented. For example, among the three search engines whose SERPs are shown in Figure 1.1, Bing and Yahoo choose to display some related image results in a separate module (other examples include video, related entities, and shopping modules). Such result modules are provided by vertical search systems that operate independent of the main web search engine, which retrieves the algorithmic search results. The vertical search systems are architecturally similar to the main web search engine, except that they operate at a smaller scale, and they often deal with structured or semi-structured databases, as data source, instead of the unstructured text content in the Web.

Finally, an important module commonly found in SERPs is the advertising module (also-called the sponsored search module). The results displayed in this module are provided by the advertisers who hope to increase the visibility of their site by attracting the attention of search engine users. Each time a user visits an advertiser's site by clicking a result link in this module, the search engine charges the advertiser a small fee. This module forms the main income source for most commercial search engine companies.

Based on what we discussed above, one can talk about three main actors in the commercial web search business: web search engines, advertisers, and end users. Each of these actors have their own set of objectives that they try to optimize. The search engine company is the primary actor in any search business since it solely provides almost the entire machinery required for running the business. Like any other company in the service sector, the main goal of a search engine is to increase its profit by monetizing the service it provides to its users. Monetization often happens through the user clicks on the advertisements as we discussed before. In the mean time, the search engine tries to keep the financial cost of running its business low. In practice, the total cost of running a web-scale search service involves the cost of operating multiple data centers, paying a large electricity bill (IT equipment, cooling system, and UPS consume the most electricity in a typical data center), maintaining vast amounts of hardware, paying a large number of employees, and spending on research and technology, among others. The advertisers are similar to the search engine in that their primary objective is monetization. However, they achieve this indirectly by increasing the volume of the web traffic flowing toward their sites from the search engine. The advertisers are often constrained in their budget and aim to reduce the amount of fees they pay to the search engine. Finally, the end users are the main beneficiaries of the web search ecosystem. Their objective is to satisfy some information need by submitting queries and consuming the results returned by the search engine. They are often constrained by the time they can allocate to searching. This, in turn, forces the search engine to provide its results with low latency.

In accordance with the classification mentioned above, we can come up with three complementary views of a search engine: search-engine-centric, advertiser-centric, or user-centric views. Although the other two are at least equally interesting, throughout the rest of the book, we adopt a search-engine-centric view. In particular, we are interested in techniques that underlie the operations of a large-scale web search engine.

As we mentioned, the main objective in running a search engine company is to create and operate a search service that will maximize the profit. As in any other business, the profit increases with increasing revenue and decreasing spendings. Increasing the advertising revenue requires attracting more advertisers, and this requires having a large base of end users who will visit the advertisers' sites. On the other hand, establishing a large and stable user base requires providing an effective search service that maintains a certain level of quality in satisfying the information needs of users. That is, the search engine is expected to provide relevant and useful results and be able to do this in a timely manner. An effective way to decrease the spendings of a search engine is to reduce its operational costs, most importantly the energy spending. Therefore, the search engine company needs to devise certain optimizations that will reduce its operational costs, potentially without hurting the quality of the provided service (most importantly, the relevance of the retrieved results and the time it takes to retrieve them).

1.2   BASIC SEARCH ENGINE ARCHITECTURE

The best way to describe the functional components of a search engine is perhaps by making an analogy to a librarian. Although this is not a perfect analogy, it serves our purposes well. In daily life, a librarian performs three basic tasks. First, they identify some important or recently published books and request a copy of them to be acquired for the library. Second, they create an index of the content of all books currently stored in the library so that the books can be easily located upon request. Third, they answer information requests coming from the clients of the library by providing the clients with pointers to the books that may contain the requested information.

At the bare minimum, a web search engine performs essentially the same three tasks. It locates the web pages that may be worth fetching from the Web and presenting in search results to the users; it prepares an easily searchable index of the fetched pages so that pages matching user queries can be quickly located; and, finally, it retrieves pages as potential answers to user queries depending on pages' estimated relevance to queries. These three tasks are known as the web crawling, indexing, and query processing tasks, respectively.

In every large-scale web search engine, there is a functionally isolated system that performs one of the tasks mentioned above. As illustrated in Figure 1.2, at the very high level, these systems operate in a pipelined fashion. The web pages are fetched by the crawling system and stored in a web repository after some filtering. The indexing system reads the pages from this repository and creates certain data structures that represent the textual content of web pages in a form that facilitate matching pages to keyword queries. Finally, the query processor evaluates the user query by processing these data structures and provides the user with a number of links (usually ten links) to matching pages. In this architecture, the crawling and indexing tasks are performed in an offline manner as they do not require any interaction with the end users. The query processing

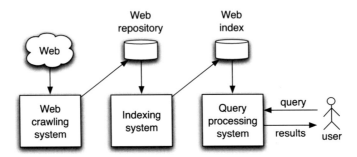

Figure 1.2: The three main tasks in a web search engine: crawling, indexing, and query processing.

task, however, can be considered as an online task since it requires interaction with the users. All three tasks face various scalability and efficiency challenges, as we will discuss in the next section.

In theory, the high-level search engine architecture we illustrated in Figure 1.2 can be deployed at different scales, ranging from a single computer to an entire data center. In practice, large-scale web search engines usually distribute their operations over multiple, geographically distant data centers. Consequently, the web repository and various data structures may be partitioned or replicated in different data centers. Within a data center, similar data and/or task splitting is performed at the level of computer clusters. These clusters are the basic units for carrying out major functional tasks, such as web crawling, indexing, and query processing. Depending on the nature of the task, the number of computers constituting these clusters can range from a few computers to tens of thousands.

1.3   SCALABILITY ISSUES

A common scalability issue affecting all three systems is the growth of the Web. Here, the term "growth" refers to two different types of growth: the growth in the amount of available web content and the growth in the user interest to search this content for information. While the former factor affects the scalability of the crawling and indexing systems, the latter factor affects the scalability of the query processing system. As we will discuss below, the scalability of each system has different implications for the success of a web search engine.

A scalable web crawling system needs to operate at the speed at which the content is created in the Web. This means that the web content should be discovered, downloaded, and stored in the repository of the crawler without much delay after it was created. Assuming enough network bandwidth and storage, a scalable crawler can continue to download an increasing number of pages without "choking" or requiring additional hardware resources, and sustaining a high page download speed, on average. Without an efficient crawler, the search engine faces the risk of turning into a library that has a limited coverage of published books. Along with growing its repository, the web crawling system should regularly refresh the content it has previously down-loaded, increasing the freshness of its repository. Going back to our analogy, this means that the search engine should act like a library that stores the most recent editions of the books.

The indexing system needs to convert, in an efficient manner, the ever-increasing number of pages in the web repository into proper index data structures. These data structures can be recre-ated from scratch in regular batches or by updating the previously created index data structures *in situ*. In either case, the generated data structures should not be excessive in size. A scalable index-ing system sustains a certain indexing speed as the size of the data structures gets bigger, without requiring additional hardware resources. An inefficient indexing system results in a search engine that resembles a library where recently acquired books cannot be located even though their copies are physically available in the library.

Finally, the query processing system needs to cope with two different issues: the increase in the size of the underlying index data structures and the increase in the query workload. In either condition, a scalable query processing system manages to provide its users with high quality search results and maintains a reasonable response latency. A search engine with a slow query processing system resembles a library with a long line of impatient clients waiting in the front desk.

As we discussed above, increasing the scalability of a search engine brings certain benefits, such as high coverage of the Web, up-to-date index data structures, and good result quality. These, in turn, result in end users who are satisfied with the search service, eventually increasing the user base of the search engine and leading to a richer advertising ecosystem. Therefore, although not directly, the scalability of a search engine can have implications for its monetization, which is the main objective for a commercial web search engine, as mentioned in Section 1.1.

But, how can we make a web search engine scalable? An intuitive answer to this question would suggest throwing more computational resources into the crawling, indexing, or query pro-cessing systems, depending on where the performance bottlenecks are identified. In many cases, this is indeed a feasible approach and is what most search engine companies do. However, the financial cost associated with this approach forms a major drawback. In practice, increasing the amount of hardware implies additional operational, maintenance, and depreciation costs for the search engine. Moreover, other constraints, such as the space limitations of data centers, may prevent the sustainability of this approach in the long term.

A potentially better approach to alleviate the scalability issues is to invest in the devel-opment of more efficient systems. In general, improving the performance of the data structures and algorithms employed in the time-critical components of a search system, when coupled with further architectural optimizations, always pays off in the long term. Moreover, this approach is more sustainable and low-cost compared to simply using more hardware. In the remaining chap-ters of the book, we focus on various performance optimizations that web search engines employ to improve their scalability and efficiency.

CHAPTER 2

The Web Crawling System

Web search engines create web repositories, essentially caching the online content in the Web on their local computer systems. These repositories provide quick access to the copies of the pages in the Web, helping to speed up the indexing and search processes. A search engine, at all times, aims to minimize the potential differences between its local page repository and the Web due to the coverage and freshness requirements we mentioned in Chapter 1. This, however, is a very challenging task since the Web is in continuous and fast evolution: a large number of web pages are created and deleted every second, while the content of the existing pages are modified. Hence, the web repositories of search engines are in a constant race to catch up with the evolving Web.

Creating a web repository would be a straightforward task if the URLs of all pages existing in the Web were somehow available to the search engine.[1] In that case, the search engine could simply iterate over the URLs and download the corresponding pages from the Web. In practice, however, the URLs of web pages must be discovered. One way to discover the URLs is to mine data sources that are likely to contain them. This kind of passive URL discovery can be performed by analyzing human curated data (e.g., Wikipedia dumps), user-generated content (e.g., tags and comments), web browsing data (e.g., toolbar and browser logs), or any other data source that may contain URLs. The major search engine companies usually have access to this kind of auxiliary data sources as they operate various communication services and provide their own web browsers.

The passive URL discovery techniques mentioned above may enable fast discovery of newly created content. However, on their own, these techniques are not sufficient for web-scale discovery due to the low coverage of URLs that could be extracted. As the primary means for URL discovery, large-scale web search engines employ a proactive technique that directly deals with

---

[1]In fact, the sitemap protocol, which will be discussed later, provides a convenient way for this kind of URL discovery, but unfortunately this protocol is not yet widely adopted by the web community.

the Web itself. This technique is known as web crawling, and the related software implementations are referred to as web crawlers.

A large-scale web crawler is responsible for two different tasks that are performed in tandem. The first task is to locate previously not discovered URLs by iteratively following the hyperlinks between the web pages. The second task is to refetch from the Web the content of pages that are currently stored in the repository. The former task aims to increase the coverage of the web repository, while the latter task aims to attain a certain level of freshness. While performing these two tasks, the crawlers are faced with a large number of scalability and efficiency challenges, mostly stemming from certain external factors. In this chapter, we review these factors and have a peek at the techniques employed in large-scale web crawlers as a remedy.

In Section 2.1, we provide an overview of a large-scale web crawling system. In Sections 2.2 and 2.3, we present the techniques used in web crawlers to improve the coverage and freshness of their web repository, respectively. The issues involved in web repository management are discussed in Section 2.4. In Section 2.5, we discuss distributed web crawling, which is an important architectural solution to achieve scalability. We present certain external factors that may hamper the performance of a web crawler in Section 2.6. A short literature survey is provided in Section 2.7. The chapter ends in Section 2.8 with a brief overview of some open research problems related to web crawling.

2.1   BASIC WEB CRAWLING ARCHITECTURE

In search engine companies, often multiple crawlers are operational at a given point in time. These crawlers may be operated by different departments that work independent of each other and may serve different purposes. For example, besides the crawler of the main web search engine, there may be crawlers running within various vertical search systems or non-production crawlers running for experimental purposes. Operating multiple web crawlers in an uncoordinated manner possesses various risks for the search engine company. These risks include the saturation of company's available network bandwidth (with potential to block some of its own web services) or overloading external websites with too many crawl requests (with potential to turn into some sort of denial-of-service attack).

As a remedy for the coordination problem discussed above, search engines usually implement a web page fetching system that is shared by multiple crawlers. This system acts as a gateway between the individual crawlers and the Web. Every time a particular web crawler needs to fetch a page from the Web, it issues a new fetching request to this system with appropriate input (e.g., the URL of the page to be downloaded). In practice, this fetching system is usually composed of two subsystems: low-level and high-level fetching systems. Individual web crawlers issue their requests first to the high-level fetching system, which then communicates those requests to the low-level fetching system. Having a shared web page fetching system relieves the burden of implementing the same functionality in different crawlers. Moreover, it provides a robust and highly

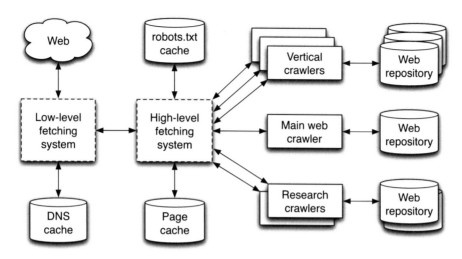

Figure 2.1: A two-level web page fetching system shared by different web crawlers that are simultaneously operating in the search engine.

tuned platform that enables efficient download of web pages. The architecture of such a system is shown in Figure 2.1.

The low-level fetching system implements the most basic network operations, such as resolving the DNS of URLs to an IP address and fetching the content of a page from the Web. Figure 2.2 illustrates the basic process that the low-level fetching system employs to download a web page. The main input to this layer is a URL, which is provided by the high-level fetching system. As the first step, the input URL is parsed to extract the domain name of the web service that stores the URL. The DNS server is then contacted to obtain a mapping from the extracted domain name to an IP address. The (domain name, IP address) mappings are stored in a DNS cache (with an expiration timestamp) to prevent repetitive accesses to the DNS server to resolve the same domain name. DNS caching significantly reduces the overhead of DNS resolution. After the IP address is obtained, the fetching system opens a TCP connection to the corresponding web server, followed by an HTTP connection. The content of the page is downloaded over this HTTP connection and is passed to the high-level fetching system. Multiple HTTP requests may be issued to a web server over a single keep-alive TCP connection.

The high-level fetching system implements various crawling mechanisms and policies that are applicable to all operational crawlers in the search engine. For example, this system implements a throttling mechanism that prevents the low-level fetching system from being flooded with too many download requests. Server-, host-, and subnetwork-level politeness constraints are also implemented by this system. These constraints safe-guard external websites from being overloaded with too many page requests concurrently issued by the crawlers. The high-level

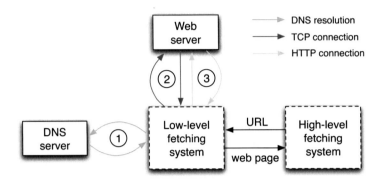

Figure 2.2: The low-level process involved in fetching a web page from the Web.

fetching system ensures that the low-level fetching system establishes only a single connection to a particular server at any given time. This policy can be implemented by assigning the requested web pages to a number of download queues in a way that all pages belonging to the same website are always added to the same download queue, and assigning each queue to a unique crawling thread. Finally, a delay is introduced between the successive requests to the same server (typically in the 10–20 s range). Such politeness constraints often conflict with the need to fully utilize the available hardware and network resources.

The high-level fetching system usually maintains a cache of recently downloaded web pages. Page caching enables the reuse of downloaded pages to answer successive requests from different crawlers, reducing the crawling workload of the low-level fetching system. The high-level fetching system also ensures that the robots exclusion protocol is obeyed by every client crawler. In this protocol, each website publishes a robots.txt file on the Web. This file includes certain instructions that will guide crawlers about how they should crawl the pages on the website. An example robots.txt file is shown in Figure 2.3(a).[2] Usually, the robots.txt file of each website is maintained in a cache (with an expiration timestamp) to eliminate the need to retrieve this file from the Web for every crawling request.

Occasionally, the robots.txt file contains a link to a sitemap file. Sitemaps are XML files created by webmasters to inform the web crawlers about the URLs served by a web server. A sitemap file can include additional meta-data about the served web pages, such as the last update time, frequency of change, and relative importance in the website. A crawler may use this information to increase its web coverage, improve the freshness of its repository, or prioritize the download of pages according to their declared importance. A sitemap often complements a robots.txt file in that it behaves as a URL inclusion protocol,[3] rather than a URL exclusion protocol. The sitemap

---

[2]More information about the robots exclusion protocol is available at http://www.robotstxt.org.
[3]More information about the sitemap protocol is available at http://www.sitemaps.org.

```
# robots.txt file

User-agent: googlebot          # all services
Disallow: /private/            # disallow this directory

User-agent: googlebot-news  # only the news service
Disallow: /                    # on everything

User-agent: *                  # all robots
Disallow: /something/          # on this directory

User-agent: *                  # all robots
Crawl-delay: 10                # wait at least 10 seconds

Disallow: /dir1/               # disallow this directory
Allow: /dir1/myfile.html       # allow a subdirectory

Host: www.example.com          # use this mirror
```

(a) An example robots.txt file.

```
<!-- sitemap.xml file -->

<?xml version="1.0" encoding="UTF-8"?>
<urlset xmlns= "http://www.sitemaps.org/schemas/
    sitemap/0.9">

<url>
<loc>http://www.test.com/</loc>
<lastmod>2015-04-03</lastmod>
<changefreq>weekly</changefreq>
<priority>0.65</priority>
</url>

<url>
...
<url>

</urlset>
```

(b) An example sitemap.xml file.

Figure 2.3: Example files illustrating the robots exclusion and sitemaps protocols.

files need to be recrawled and cached as in the case of the robots.txt files. An example sitemap file is shown in Figure 2.3(b).

Given the availability of the low-level and high-level fetching systems, an individual web crawler's responsibility is reduced to perform two complementary tasks: increasing the web coverage by discovering/downloading new pages and refreshing the web repository by refetching already downloaded pages. To this end, the web crawler maintains two separate (logical) download queues containing URLs, one for the discovery task and one for the refreshing task. The main algorithmic challenge for web crawler developers is to devise techniques for prioritizing the URLs in these queues. We discuss possible techniques in the following two sections.

2.2   EXTENDING THE WEB REPOSITORY

Web crawlers rely on the hyperlinks between web pages to discover new URLs. Therefore, it may be useful to first present the hyperlink structure of the Web before discussing the URL discovery process employed by the crawlers. The web pages and the underlying hyperlink structure can be viewed as a directed graph. In this representation, each vertex in the graph corresponds to a web page. If a web page provides hyperlink(s) to another page, a directed edge exists between the corresponding vertices in the graph. A crawler essentially discovers new pages by traversing the vertices connected by the edges in the graph. When the crawler starts operating for the first time, it needs some seed web pages that will act as entry points to the web graph. These pages are usually selected from so-called hub pages, which point to a large number of potentially important web pages.

Essentially, the discovery process splits the pages in the Web into three disjoint sets. The first set includes the web pages whose content is already downloaded by the crawler. This set corresponds to the web repository of the crawler. The second set contains the web pages whose URLs are discovered (by following the hyperlinks pointing to these pages), but whose content is not yet downloaded. This set is referred to as the frontier of the crawler. Finally, the third set includes the web pages whose URLs are not discovered yet. These three sets are illustrated in Figure 2.4.

In some early web crawler implementations, the discovery was performed as a batch task that is carried out in separate sessions. In each crawling session, the web pages were downloaded for a certain time period or until the number of pages downloaded in the session reaches a predetermined threshold. Session-based web crawling was motivated by the scalability challenges faced in the early web crawlers. As we will discuss shortly, throughout the web crawling process, certain data structures continue to grow. At some point, the system resources become insufficient for healthy execution of the crawler. Session-based web crawling prevents the crawler from "choking" by cleaning the data structures used in the crawling process each time before starting a new crawling session.

In large-scale web search crawlers, the discovery is an incremental process. At each iteration, the crawler selects a target URL from the URLs currently available in its frontier and passes this

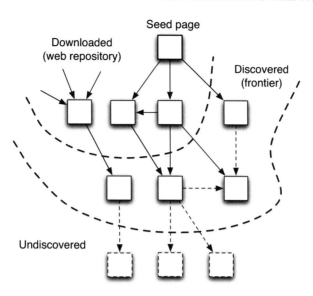

Figure 2.4: The URL discovery process splits the Web into three disjoint sets.

URL to the high-level fetching system for download. The HTML content returned by the high-level fetching system is passed to the repository management module, which decides whether the content should be stored in the repository or not. The content is then passed to the parsing module for further processing. The parsing module performs certain extraction tasks, the most important being the extraction of the URLs contained in the page. Every extracted URL goes through a normalization process, where the case-insensitive components of the URL (e.g., the scheme and host components) are converted to lowercase and the default port number, if present, is removed from the URL. Moreover, the "www" prefix may be removed from or added to the domain name if either version is determined to redirect to the other. Further normalization may include the addition of a trailing "/" character to the URL or conversion of relative paths to absolute paths, among others (some of these normalizations may not preserve the semantics of the URL). The normalized URLs are then looked up in a data structure that maintains all URLs seen by the crawler so far. This lookup is known as the URL-seen test. If the extracted URLs are not found among the seen URLs, they are added to the frontier of the crawler. They are also added to the data structure that stores the seen URLs. The process continues with another URL selected from the frontier of the crawler. The entire process is depicted in Figure 2.5.

The data structures used during the discovery process are updated in an incremental fashion and keep growing as new pages are downloaded and links are discovered. Therefore, the efficiency of these data structures plays an important role in the scalability of the crawler. Given the vast number of URLs that need to be maintained, using only in-memory data structures is not feasible.

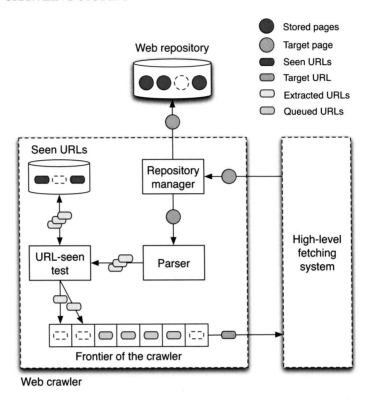

Figure 2.5: The high-level mechanism employed by a crawler to discover and download new web pages.

The data structure that keeps the discovered URLs may be implemented as a very large hash table which is mostly stored on disk and partially cached in the memory. Despite the high cache hit rates, this is still a naive implementation that may turn the disk into a bottleneck due to the extremely large number of accesses made to the hash table during the URL-seen tests. A more scalable implementation is based on accumulating the seen URLs in multiple on-disk buffers, which are periodically merged with a sorted, on-disk array that stores all URLs seen so far.

Another implementation issue is about the order in which the pages will be downloaded from the Web. A simple option here is to download the URLs in a random order, disregarding the properties of web pages. A more commonly adopted alternative is to download the URLs in the order they are discovered by the crawler, resulting in a breadth-first crawling strategy. Both of these options are easy to implement (e.g., using an on-disk FIFO queue). However, these implementations ignore the fact that, from the perspective of the search engine, some web pages are more valuable than the others. In practice, commercial web crawlers employ certain URL priori-

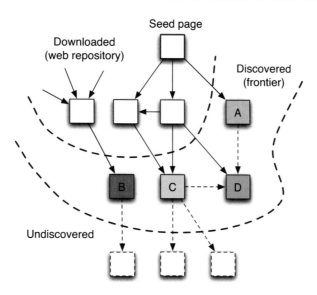

Figure 2.6: An example illustrating different URL prioritization strategies. In this example, the frontier of the crawler includes four pages: A, B, C, and D (more intense red color indicates higher link quality). A crawler that employs breadth-first, link-quality-based, or in-degree-based URL prioritization strategy would first download page A, B, or C, respectively. Note that page D, which has three inbound links, would not be selected by the in-degree strategy since, at this point, only one of the inbound links has been discovered by the crawler.

tization strategies to download important pages earlier, as an attempt to form a high-quality web repository. In general, there are two complementary lines of approach to measure the importance of pages. The approaches in the first line exploit the connectivity of pages in the web graph. A simple approach, known as the in-degree measure, estimates the importance of a page based on the number of inbound links of the page. The well-known PageRank measure goes one step further and incorporates the importance of linking pages into the computation, i.e., the importance of a page is assumed to be proportional to the number and importance of other pages linking to the page. Essentially, the PageRank measure computes a probability distribution showing the likelihood of a user randomly surfing the Web to arrive at a particular page. The second line of approach is more direct in the way it captures the importance of a page. In this case, the importance is measured by the potential impact the page would make on the search result quality or the users' engagement with the search engine. The web-centric and user-centric importance measures can be combined into a single importance measure by proper weights. We illustrate several URL prioritization strategies in Figure 2.6.

	A	B	C	D
PageRank	0.0003	0.0007	0.0002	0.0001
Average daily click count	47	332	2	1974
Last download time	2 hours ago	1 day ago	8 days ago	6 hours ago
Estimated update frequency	daily	never	minutely	yearly

Figure 2.7: An example illustrating different URL prioritization strategies for refreshing the web repository. In this example, there are four pages in the web repository: A, B, C, and D. The refreshing strategy that is based on link quality would first refresh page B, which has the highest PageRank, while the strategy based on search impact would first refresh D, which has the highest observed average daily click count. The age-based refreshing strategy would refresh page C first since it is the least recently downloaded page in the repository. Finally, the longevity-based refreshing strategy would potentially refresh page A first since pages B and D have too high longevity, while page C has too low longevity.

Typically, the URLs in the frontier of the crawler are stored in a disk-based priority queue. In general, importance-based URL prioritization leads to complications in the implementation of this priority queue. A standard heap implementation is not efficient as the insertion cost of URLs is not constant time. A more practical solution is to assign the URLs into multiple FIFO queues such that URLs with similar estimated importance values fall into the same queue. When URLs are scheduled for download, they can be sampled from each queue at a rate proportional to the average importance of the URLs in the queue.

2.3 REFRESHING THE WEB REPOSITORY

The URL-seen test mentioned in the previous section constrains every web page to be downloaded at most once. This implies that, as the content of the pages in the Web continues to change, some of the pages in the web repository become stale. That is, they are not content-wise identical to their originals in the Web anymore. Not having relatively recent versions of the pages may lead to degradation in the result quality of the search engine. Crawlers cope with this problem by selectively refetching the pages in their repository over time. This process is known as refreshing. The refreshing decisions are not trivial because it is not possible for a crawler to determine precisely, before downloading a page, whether the page's content has changed or not. Deciding not to refresh a modified page leads to staleness, while refreshing an unmodified page wastes resources.

As in the case of the discovery process, the most naive refreshing technique is to select the pages to be refreshed in a random manner (or with uniform probability). A better alternative is to refresh important pages (e.g., those with high link-quality scores or with high impact on search quality) relatively more often. A more direct approach is to reduce the average staleness of pages by first refreshing pages with the highest age (the age of a page is defined as the time passed since the page was last downloaded by the crawler). Yet another alternative is to prioritize pages according to their longevity, i.e., the estimated update frequency of the original page. In this case, the crawler prefers to refresh web pages with medium longevity. The intuition here is that high-longevity pages (e.g., those that do not change over a long period of time) do not require being refreshed very often, while refreshing low-longevity pages (e.g., those that change many times in a short time period) is not very useful as these page would turn stale soon after they are crawled anyway. Figure 2.7 illustrates several refreshing strategies on an example. In practice, the quality-based refreshing strategies need to be relaxed or combined with other strategies since low-quality pages may never be refreshed, leading to a starvation problem.

Refreshing may be implemented in a way similar to importance-based URL prioritization, which we discussed in the previous section: a number of disk-based FIFO queues can be used to support scheduling of pages for refreshing. In this case, however, each page is assigned to a queue according to the estimated utility of refreshing the page. Once a page is removed from the head of a queue and refreshed, it is inserted back to the tail of the same queue. The queues can be periodically recreated to reflect the potential changes in the estimated utilities of pages.

2.4   MANAGING THE WEB REPOSITORY

The web pages downloaded by the crawler are stored in a web repository. The repository maintains only the most recently crawled versions of the pages, i.e., refreshing a web page overrides the previous content in the repository. In this respect, web crawler repositories are different from archival systems, which keep different versions of content that evolves in time. A web repository usually stores only the raw HTML pages, not their text. Most web pages are amenable to compression due to the high repetition in HTML tags and the textual content of the page. Therefore, the pages are stored in a compressed form, saving storage space.

The storage system provides mechanisms that facilitate both random and bulk access to the pages in the web repository. The former type of access facilitates reading or updating individual web pages. The latter type of access is required since the complete web repository needs to be read, regularly, to build an inverted index and some other data structures (see Section 3.1). As both types of data access are basic, the content can be stored on the flat file system, instead of a structured database. Each page can be accessed through an identifier that is obtained by mapping the URL of the page to a hash value. The hashing is not free of collisions, but the likelihood of having two distinct URLs represented by the same hash value is extremely low. Moreover, collisions do not create a problem since, upon a hash value match, the actual content of the page is checked as well. Certain information about each page, such as the location of the page on disk, its size, and a history of HTTP status codes and download times, are maintained in a data structure that acts like a catalog. This data structure may be implemented as a B+ tree.

Due to the vast size of the web repository, the backend storage system is distributed over many nodes (e.g., a cluster of computers or an array of network drives). The pages can be distributed on the nodes uniformly or based on hash value ranges. The mapping between the web pages and storage nodes is maintained in an index. Hash-based distribution leads to a much smaller index since only a mapping from the hash value ranges to node identifiers needs to be stored. However, uniform distribution is more practical since it simplifies the redistribution of pages among the nodes when new nodes are added to the storage system. It also simplifies the implementation of fault tolerance mechanisms.

Finally, the stored web pages may expire and eventually be purged from the repository. The reasons for expiration include consecutive server errors while attempting to fetch the page or editorial flagging (e.g., due to spam or malware content). The meta-data associated with purged web pages may be kept in the system for reporting purposes after the page content is deleted.

2.5   DISTRIBUTED WEB CRAWLING

Obtaining a high coverage of the Web and maintaining the freshness of the web repository require sustaining a high page download speed, which is the most important efficiency objective for a crawler. If a crawler can download the pages faster, it can cope better with the growth and evolution of the Web. Hence, it can achieve a higher web coverage and page freshness, perhaps with some positive impact on the search quality as well.

In its simplest form, the low-level fetching system can be implemented as a single-thread that fetches only one web page at a given point in time. Obviously, due to its sequential nature, this type of crawling does not scale well in terms of the download speed. The speed at which the pages are downloaded can be increased simply by running more fetcher threads and downloading the pages concurrently. Typically, a single crawling node can accommodate around a hundred fetcher threads, each handling an HTTP connection to a different web service. Using more threads results in too much context switching overhead in the CPU and does not bring additional speed increase.

For an efficiently implemented web crawler, the main bottleneck is the network bandwidth available to the crawler. Usually, multi-threading is not sufficient to saturate the network bandwidth if few crawling nodes are used. Therefore, large-scale web crawlers employ large clusters containing many nodes (e.g., a few hundred nodes), trying to utilize the available network bandwidth as much as possible. Given a robust crawler that runs on a single node, it is relatively easy to build a parallel crawler that runs on multiple nodes. In an extreme case, the crawling nodes can run in an embarrassingly parallel mode, downloading the web pages independent of each other. This approach is easy to implement, but it results in high redundancy as the same web page may be downloaded, repetitively, by different nodes.

A possible solution to eliminate the redundancy issue is to partition the space of URLs among the crawling nodes. The URL space can be partitioned by mapping each URL to a hash value and then assigning evenly split hash value ranges to the crawlers. This way, each crawling node becomes responsible for fetching the URLs whose hash values are assigned to itself only.

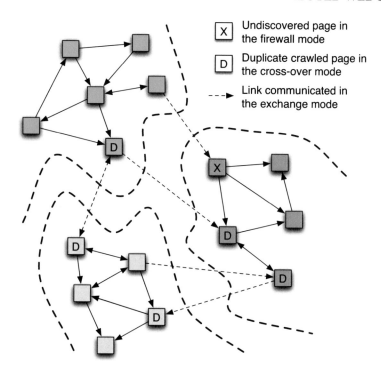

	Undiscovered page in the firewall mode
X	
D	Duplicate crawled page in the cross-over mode
- - →	Link communicated in the exchange mode

Figure 2.8: Coordination issues in distributed web crawling illustrated by a web graph partitioned over three crawling nodes.

Given a uniform hashing function, the crawling workload of the nodes is expected to be balanced. This solution, known as the firewall mode, guarantees that each page will be downloaded at most once. However, the lack of coordination in the firewall mode implies that hyperlinks whose source and destination pages are assigned to different nodes cannot be followed by any node. This, in turn, leads to reduced web coverage as some web pages are never discovered. We illustrate this issue in Figure 2.8, which shows a web graph partitioned over three crawling nodes indicated by different colors. Page X, which is assigned to the blue crawling node, has an inbound link from a page assigned to the green crawling node. This link is not followed by the green crawling node because it is not responsible for downloading X. Since X does not have any inbound link from a blue page, it will never be discovered, reducing the coverage of the crawling system.

A solution to the discoverability problem discussed above is to follow a hyperlink (only one hop) even though its destination URL is assigned to another node. In our example, the green crawling node would follow the link to page X and would be able to discover it. This technique, known as the cross-over mode, solves the coverage issue in the firewall mode. However, it reintroduces the redundancy issue as some pages may now be downloaded multiple times by different nodes. In our example, under the cross-over mode assumption, all of the pages labeled as D would be downloaded twice since they have links from pages assigned to two different crawling nodes.

The solution, known as the exchange mode, eliminates the coverage and redundancy issues mentioned above. In the exchange mode, the discovered non-local links are communicated to the crawling nodes that are responsible for fetching them. In Figure 2.8, the links to be exchanged between two crawling nodes are denoted with dashed lines. The volume of hyperlinks that need to be communicated over the network can be significantly reduced if the web pages are partitioned among the nodes based on their domain names instead of URLs. This is because websites form a natural clustering of links, and thus domain-based partitioning reduces the number of cross-node links. Another possible optimization is to communicate the URLs, periodically, in small batches rather than communicating them immediately upon their discovery. This prevents the same URL from being communicated multiple times in a short time period. Similarly, caching frequently communicated URLs may help to reduce the volume of link exchanges between the nodes. The exchange mode is the common practice adopted in web crawlers due to its simplicity and guarantees on web coverage and crawling workload.

A natural extension to parallelizing the crawling process on a cluster of nodes is to distribute the process over data centers that are geographically far away. In this case, the websites can be partitioned across the data centers taking into account the network proximity between the websites and the data centers. This way, since the propagation delay and potential routing delays are reduced, it may be possible to achieve higher download speeds when fetching pages. Moreover, compared to the centralized crawling architectures, geographically distributed web crawling offers improved fault tolerance due to its resilience to network partitions. Some large-scale search engines implement this kind of multisite systems and crawl the Web from different data center locations.

2.6 FACTORS AFFECTING CRAWLING PERFORMANCE

The Web poses various obstacles that can negatively affect the performance of web crawlers. Some of these obstacles stem from the malicious intent of website owners. Well-known examples include the delay attacks, spider traps, and link farms. Although there is no malicious intent behind them, certain other factors may also affect the crawling performance (e.g., website mirrors and soft 404 pages). In this section, we briefly review these external factors.

A web server is said to perform a delay attack if it introduces artificial time delays in its responses to crawlers' requests. These delays cause the crawling threads to become idle for a time period longer than usual. If too many fetching threads become the subject of such delay attacks, the average download speed of the crawler may decrease significantly. Web crawling systems can cope with delay attacks by setting an upper bound on the waiting time of the crawling threads. Moreover, web servers that have high response latency are usually crawled less frequently.

A spider trap is a web server that dynamically generates pages containing many useless links to other dynamically generated pages, essentially creating an infinite loop for the crawlers.

Spider traps may create a lot of useless workload for a crawler, creating an important hazard for its performance. Spider traps are usually very difficult to detect and avoid.

Another important performance hazard for crawlers are link farms. A link farm is essentially a large group of websites that provide artificially created links among the websites in the group. By means of these links, the link farm aims to boost certain link-based importance scores computed in web search engines, thus hoping to move the pages in the link farm to higher ranks in search results. Although the main objective of link farms is search engine optimization, they also harm crawlers since a crawler may allocate a significant portion of its resources to download the web pages in link farms.

In website mirroring, a large number of web pages are stored on one or more websites that are accessible by different domain names. Website mirrors waste crawling resources since the crawler may end up fetching the same page content redundantly from different mirrors. Therefore, early detection of mirror sites is important. Website mirrors can be detected by analyzing the link structure of websites and measuring the similarity of their content.

When a web server cannot find the requested page, it is expected to return an "HTTP 404 Not Found" response code indicating the unavailability of the page on the server. However, instead of this standard response code, some web servers prefer to return a custom-designed error page to indicate that the requested page is not available on the website, together with an "HTTP 200 OK" response code indicating that the page was loaded properly. These so-called soft 404 error pages can degrade the performance of a web crawler since they are usually not worth downloading and indexing. A common soft 404 page detection technique is to generate arbitrary URLs that are unlikely to exist on any website and request these URLs from the web servers to learn the custom soft 404 message patterns used by each website.

2.7   LITERATURE ON WEB CRAWLING

In the literature, there are a number of papers that describe the issues encountered in designing a web crawler. The details of the Internet Archive's web crawler were presented in Burner [1997]. In this multi-process crawler, each crawling process is responsible for crawling a fixed set of websites statically assigned to itself and maintains a download queue for pages belonging to those sites. Cross-site links are first written to the disk, and then periodically moved to the corresponding download queues, filtering out duplicate links. Google's web crawler was briefly introduced in Brin and Page [1998]. This crawler design is centralized in the sense that the URLs are assigned to different crawling processes by a central URL server. Although the paper touches certain performance issues, such as DNS caching, it does not disclose most design details of the crawler. A clear and detailed design for a multi-threaded crawler, called Mercator, can be found in Heydon and Najork [1999]. This paper elaborates in great detail on many design challenges and the implementation of the data structures employed as solution, including those for politeness requirements and URL-seen tests. A distributed crawler, called PolyBot, was presented in Shkapenyuk and Suel [2002]. The design of this crawler is modular as different tasks are split

Table 2.1: A list of open source web crawlers

Crawler	Description
BUbiNG	Distributed crawler (GNU GPLv3+)
GRUB	Distributed crawler (GNU GPLv2)
Heritrix	Internet Archive's crawler (Apache license)
Norconex HTTP Collector	Multi-threaded crawler (Apache license)
Nutch	Distributed crawler with Hadoop support (Apache License 2.0)
PHP-Crawler	Script-based crawler (BSD license)
Scrapy	Crawling framework (BSD license)
Wget	Computer program to retrieve pages (GNU GPLv3+)

between two types of modules referred to as the crawling system (responsible for DNS caching, politeness, robots exclusion protocol) and the crawling application (responsible for URL prioritization). In this respect, the design is similar to the shared page fetching system we described in this book. In Boldi et al. [2004], the design details of a research crawler, called UbiCrawler, were presented. Along with engineering some solutions to common scalability issues, the work discusses the techniques for improving the fault tolerance of a web crawler. IRLbot is a more recently developed web crawler, whose design was presented in Lee et al. [2008]. This crawler implements novel data structures and algorithms that render it extremely scalable and fast. The scalability improvements achieved in this design are mainly due to an efficient implementation of the URL-seen test and the techniques that reduce the overhead of crawling large link farms. The relatively recent work in Boldi et al. [2014] presents the design of BUbiNG, a scalable, distributed crawler, which was built based on the experience obtained throughout the development of UbiCrawler. A number of open source web crawlers are listed in Table 2.1.

The work in Dasgupta et al. [2007] investigated the techniques for quicker discovery of new web content. The search impact of two different passive URL discovery techniques, relying on social bookmarking data and web browsing data (obtained from toolbar logs), were described in Heymann et al. [2008] and Bai et al. [2011], respectively. The experiments in Cambazoglu et al. [2009] indicated that faster URL discovery may have a positive effect on the quality of search results. In Lefortier et al. [2013], a technique was proposed for timely discovery of ephemeral pages, which are of interest to users for a relatively short period of time.

The performance of breadth-first, in-degree-based, and PageRank-based URL prioritization strategies were compared in Cho et al. [1998]. Prioritization of pages based on their PageRank scores was found to be better for early discovery of important URLs, independent of whether the quality of the crawled collection is measured by the in-degree or PageRank measures. The inherent tendency of breadth-first crawling to discover important pages early was demonstrated in Najork and Wiener [2001] with a large-scale crawling experiment. The work in Baeza-

Yates et al. [2005] proposed prioritizing the URLs in websites that have large numbers of URLs that are discovered but not yet crawled. This strategy was shown to perform better than the three techniques explored in Cho et al. [1998], arguing that they lead to a politeness bottleneck toward the end of the crawl. The prioritization approach adopted in Pandey and Olston [2008] considered the potential impact on search quality. In this approach, the search impact of a page was measured by the likelihood that the page will become visible to the users in the top search results. The pages were prioritized according to a linear combination of their PageRank scores and the estimated relevance of their URLs and anchor text to a sample of search queries. A hybrid URL prioritization approach, also focusing on the search impact, was proposed in Tran et al. [2015]. In this case, the search impact of a page was measured by the number of times the page was viewed or clicked in the top search results. The URLs were prioritized according to a technique that incorporates the click-through information associated with the previously crawled pages into the standard PageRank computation. Along the same line, the work in Liu et al. [2011] exploited the past web browsing behavior of users to prioritize the URLs and observed the impact on the search quality. Finally, several URL prioritization strategies were compared in Fetterly et al. [2009].

The dynamics of web change were investigated in several studies [Adar et al., 2009, Fetterly et al., 2004b, Ntoulas et al., 2004]. These studies pointed out the rapid change in the web content as well as the link structure, demonstrating the importance of refreshing web crawler repositories. Early papers [Cho and Garcia-Molina, 2003, Edwards et al., 2001] used the observed update frequency of a page as a proxy for its actual change frequency and suggested refreshing frequently updated pages more often. The succeeding work [Barbosa et al., 2005] coupled the update history with features extracted from the content of the page to predict the change. This technique was useful especially when a page had little or no update history. Other work [Cho and Ntoulas, 2002, Radinsky and Bennett, 2013, Tan and Mitra, 2010] incorporated the feedback obtained from the update history of related pages when making predictions about the freshness of a page. In Wolf et al. [2002], the authors devised a refreshing strategy to reduce the average staleness of crawled pages as well as the user frustration due to the inclusion of stale pages in search results. Along the same line, the work in Pandey and Olston [2005] proposed refreshing pages in decreasing order of their contribution to the search quality. Finally, the work in Olston and Pandey [2008] suggested focusing on the longevity (the life span of content fragments that exist in web pages over time) and proposed techniques that avoid refreshing fast-changing content as much as possible.

Publications on web repository management are relatively scarce. An early work on managing web repositories can be found in Hirai et al. [2000]. Further information about web repository management techniques can be obtained from Arasu et al. [2001]. The study in Ferragina and Manzini [2010] provides a performance comparison of different compression techniques in the context of web pages.

Distributed web crawling was considered before at different granularities. The design trade-offs in parallel crawlers was first investigated in Cho and Garcia-Molina [2002]. This paper provides important insights about the issues involved in parallel web crawling, such as web partitioning and coordination of the crawling processes. Peer-to-peer crawling systems were proposed in Suel et al. [2003] and Singh et al. [2004]. The work in Cambazoglu et al. [2008] investigated the feasibility of crawling the Web, concurrently, from data centers located in different continents. The reported empirical results indicate a potential increase in the aggregate download speed of a multisite distributed crawler with respect to a centralized crawler. In Exposto et al. [2008], a graph partitioning technique was proposed to assign the URLs to crawling sites in a multisite distributed web crawler.

Several heuristics for in-memory caching of seen URLs were evaluated in Broder et al. [2003b]. This kind of caching was shown to improve the efficiency of URL-seen tests significantly. An early work discussing the ethical issues involved in crawler design is available in Eichmann [1995]. In Brin and Page [1998], the politeness issues caused by the Google crawler were demonstrated with examples. In the literature, various techniques were proposed to detect the website mirrors [Bharat et al., 2000, Cho et al., 2000], link farms [Gyöngyi and Garcia-Molina, 2005], and soft 404 pages [Lee et al., 2009]. The technique introduced in Webb et al. [2008] allowed the detection of web spam by analyzing HTTP sessions, even before the content of spam pages are downloaded by the crawler. Techniques for detecting duplicate pages solely based on the analysis of the URL string are available in Bar-Yossef et al. [2007] and Koppula et al. [2010]. An interesting study on sitemaps can be found in Schonfeld and Shivakumar [2009].

Besides the general-purpose web crawlers used in search engines, there are web crawlers designed for specific tasks. For example, hidden web crawlers aim to discover content that is not accessible by following an explicitly available link structure [Raghavan and Garcia-Molina, 2001]. Such content includes dynamically generated pages, private sites, unlinked pages, or scripted content. If a query interface is available on the website, the content hidden in the site can be crawled (or better scraped) by issuing synthetic queries through this interface. New queries can be generated by extracting terms from the content of the retrieved pages [Ntoulas et al., 2005]. Another issue here is the automated discovery of hidden web entry points [Barbosa and Freire, 2007]. In Baeza-Yates and Castillo [2007], hidden crawling models inspired by the users' web browsing behavior were proposed.

Focused crawlers are another form of special-purpose crawlers [Chakrabarti et al., 1999]. These crawlers aim to discover content relevant to a specific topic of interest [Diligenti et al., 2000] or aspects, such as genre [De Assis et al., 2009], opinion [Vural et al., 2012], or geolocation [Ahlers and Boll, 2009], among others. The main challenge in focused crawling is to guide the discovery process toward pages that are related to a particular aspect so that they can be downloaded earlier. The relevance of a URL is usually estimated using features extracted from the linking pages and anchor text since the content of the page is not yet available. Finally, there are web crawlers that are designed to crawl websites with a specific content structure, such as web forums [Cai et al., 2008].

In Table 2.2, we provide a list of the web crawling papers surveyed in this section. A more comprehensive survey of web crawling systems can be found in Olston and Najork [2010]. Although it is slightly outdated, some sections of Arasu et al. [2001] also provide a good overview of web crawlers.

2.8   OPEN ISSUES IN WEB CRAWLING

We previously mentioned that discovery and refreshing are two tasks that are carried out separately by web crawlers. That is, there are two logical priority queues that are managed independent of each other, one for discovery and one for refreshing of the pages. The rate at which the pages are selected from these two queues has implications for the coverage and freshness of the search results presented to the users. Ideally, the pages should be selected from the two queues at different rates, according to the relative impact of the discovery and refreshing tasks on the search quality. In most commercial web crawlers, these rates are determined in an ad-hoc manner since there is no known technique to set them in an automated manner. Finding the right balance requires a thorough analysis of the trade-off between the utilities of the discovery and refreshing tasks.

Another open problem is to investigate the feasibility of the crawling techniques where URLs are discovered by external agents and communicated to the crawler. An extreme example of this kind of push-based web crawling is to place software/hardware crawling agents on the routers and ISPs in the network. These agents can continuously inspect the network packets passing through the routers to discover the URLs that users are browsing. This may enable the discovery of content that is not easy to discover by the standard web crawling techniques that rely on following links. Examples of such content are loosely connected pages and pages that have no inbound links. Moreover, using such a technique, it may be possible to discover newly created web pages much faster. Obviously, this kind of URL discovery techniques should always be designed with potential privacy issues in mind (e.g., encrypted HTTP requests/responses should be omitted).

Finally, we believe that geographically distributed web crawling needs further research attention. In this kind of crawling, the main objective is to partition the websites among the data centers such that the download speed of the crawling system is increased relative to a centralized crawler, due to the increased proximity between the crawlers and web servers. The research needs to identify effective techniques for assigning the websites to the data centers, taking into account the potential overheads due to the link exchange and distributed indexing as well.

Table 2.2: A summary of the literature on web crawling

Topic	Reference	Short description
General overview	[Olston and Najork, 2010]	Comprehensive survey on crawling
	[Arasu et al., 2001]	Overview of crawling and repository management

Crawling system design	[Eichmann, 1994]	RBSE crawler
	[Burner, 1997]	Internet Archive's crawler
	[Brin and Page, 1998]	Google's crawler
	[Heydon and Najork, 1999]	Mercator crawler
	[Shkapenyuk and Suel, 2002]	PolyBot crawler
	[Boldi et al., 2004]	UbiCrawler
	[Lee et al., 2008]	IRLbot crawler
	[Boldi et al., 2014]	BUbiNG crawler
URL discovery	[Dasgupta et al., 2007]	Techniques to increase the discovery rate of pages
	[Heymann et al., 2008]	Passive discovery through social bookmarking
	[Bai et al., 2011]	Passive discovery using web browsing data
	[Cambazoglu et al., 2009]	Search impact of faster URL discovery
	[Lefortier et al., 2013]	Discovery of ephemeral content
Frontier prioritization	[Cho et al., 1998]	Compares three connectivity-based techniques
	[Najork and Wiener, 2001]	Evaluates the breadth-first crawling quality
	[Baeza-Yates et al., 2005]	Compares several techniques
	[Pandey and Olston, 2008]	Combines PageRank and relevance to queries
	[Tran et al., 2015]	Combines PageRank and click feedback
	[Liu et al., 2011]	Exploits the web browsing behavior of users
	[Fetterly et al., 2009]	Evaluates the search impact of techniques
Refreshing web pages	[Adar et al., 2009]	Dynamics of web change
	[Fetterly et al., 2004b]	Dynamics of web change
	[Ntoulas et al., 2004]	Dynamics of web change
	[Cho and Garcia-Molina, 2003]	Based on the historical change frequency
	[Edwards et al., 2001]	Based on the historical change frequency
	[Barbosa et al., 2005]	Based on the content and historical change frequency
	[Cho and Ntoulas, 2002]	Based on the change in the site
	[Radinsky and Bennett, 2013]	Based on the change in similar pages
	[Tan and Mitra, 2010]	Based on the change in the same page cluster
	[Wolf et al., 2002]	Aims to reduces the staleness and user frustration
	[Pandey and Olston, 2005]	Based on the impact on search quality
	[Olston and Pandey, 2008]	Based on the longevity
Web repository management	[Hirai et al., 2000]	Web repository management
	[Arasu et al., 2001]	Web repository management
	[Ferragina and Manzini, 2010]	Techniques for compressing web pages
Distributed web crawling	[Cho and Garcia-Molina, 2002]	Parallel web crawling
	[Suel et al., 2003]	Peer-to-peer web crawling
	[Singh et al., 2004]	Peer-to-peer web crawling
	[Cambazoglu et al., 2008]	Multisite web crawling
	[Exposto et al., 2008]	URL partitioning for multisite web crawlers

Other issues	[Broder et al., 2003b]	In-memory URL caching
	[Eichmann, 1995]	Ethical issues in web crawler design
	[Cho et al., 2000]	Web mirror detection
	[Bharat et al., 2000]	Web mirror detection
	[Gyöngyi and Garcia-Molina, 2005]	Study on link farms
	[Lee et al., 2009]	Soft 404 page detection
	[Webb et al., 2008]	Online web spam detection
	[Bar-Yossef et al., 2007]	Duplicate page detection by URL string analysis
	[Koppula et al., 2010]	Duplicate page detection by URL string analysis
	[Schonfeld and Shivakumar, 2009]	Study on sitemaps
Hidden web crawlers	[Raghavan and Garcia-Molina, 2001]	Seminal work on hidden web crawling
	[Ntoulas et al., 2005]	Crawling the hidden web using keyword queries
	[Barbosa and Freire, 2007]	Discovering hidden web entry points
	[Baeza-Yates and Castillo, 2007]	Hidden web crawling models
Focused web crawlers	[Chakrabarti et al., 1999]	Seminal work on focused web crawling
	[Diligenti et al., 2000]	Topic-focused web crawling
	[De Assis et al., 2009]	Genre-focused web crawling
	[Vural et al., 2012]	Sentiment-focused web crawling
	[Ahlers and Boll, 2009]	Geolocation-focused web crawling
	[Cai et al., 2008]	Vertical web crawler

CHAPTER 3

The Indexing System

The indexing system, besides indexing, is responsible for a number of information extraction, filtering, and classification tasks. Here, we overload the term "indexing" to include the tasks related to the processing of web pages as well. This system provides meta-data, metrics, and other kinds of feedback to the crawling system (e.g., various link analysis measures that are used to guide the crawling process) as well as the query processing system (e.g., some query-independent ranking features). It manipulates the output of the crawling system and the query processing system operates on its output. In this respect, the indexing system acts as a bridge between the crawling and query processing systems.

One of the important tasks performed by the indexing system is to convert the pages in the web repository into appropriate index structures that facilitate searching the textual content of pages. These index structures include an inverted index together with some other auxiliary data structures, which are processed each time a query is evaluated. In practice, the time cost of processing these data structures is the dominant factor in the response latency of a search engine. Moreover, these data structures are usually retained in memory, avoiding costly disk accesses as much as possible. Therefore, the compactness and efficient implementation of the index structures is vital. Finally, the index data structures need to be kept up to date by including newly downloaded pages so that the search engine can guarantee a certain level of freshness in its results. This is achieved either by periodically rebuilding the index structures or incrementally reflecting the changes in the web repository to the index structures.

Another important objective of the indexing system is to improve the quality of the results served by the search engine. To this end, the indexing system performs two complementary types of task. The first type of task aims to eliminate the useless or harmful content that may leak into the search results presented to the users. Spam pages, malware, duplicates, and uninformative content are tried to be detected and filtered out before they are indexed and made searchable. The second type of task aims to understand the content of web pages and enrich them with certain meta-data by employing natural language processing techniques and semantic analysis. A rich set of features are extracted from the content of web pages, the web graph, query logs, and other sources of information. These features are later used by the query processing system to improve the search quality when pages are ranked against queries.

The rest of the chapter is organized as follows. In Section 3.1, we provide an overview of a typical indexing system used in web search engines. The inverted index data structure is described in Section 3.2. In Section 3.3, we illustrate various compression techniques that are used to create more compact inverted indexes. Section 3.4 presents different techniques used for building an inverted index. We discuss the techniques used for maintaining the freshness of an inverted index in Section 3.5. Section 3.6 describes two alternative strategies for partitioning an inverted index over a distributed system. In Section 3.7, we survey the literature on indexing. Finally, in Section 3.8, we provide some open research issues related to the indexing systems.

3.1   BASIC INDEXING ARCHITECTURE

Large-scale indexing architectures include a number of document processing pipelines composed of many small software modules. These modules are specialized in applying various parsing, extraction, filtering, and classification tasks on the content of web pages. Every page in the web repository goes through some of these pipelines. Each module in a document processing pipeline processes the content received from the previous module, potentially modifying or enriching it, and passes its output to the next module. Some modules may drop certain types of content (e.g., binary files, soft 404 pages, or spam pages), preventing further processing. Some important document processing pipelines are illustrated in Figure 3.1, highlighting a small number of selected modules.

Although it is an offline step, the efficiency of the modules in the document processing pipelines is important. A pipeline containing an inefficiently implemented module not only slows down the rest of the indexing system, but may also create a bottleneck for the crawling system. This is because if the pages cannot be processed at the speed they are crawled from the Web, high-speed web crawling becomes less valuable. In an extreme case, the web crawler may start dropping some of the web pages in its frontier, rather than crawling them, to match the speed of the document processing pipeline.

In general, a large-scale indexing system maintains or periodically rebuilds the following four data structures: web graph, forward index, page attribute file, and inverted index. Besides these data structures, some auxiliary data structures are also created to facilitate specific search tasks. As an example, a compressed data structure may be constructed to enable quick vocabulary search when auto-completing user queries. Herein, we discuss only the main data structures mentioned above.

The web graph, as described in Section 2.2, represents the link structure between the web pages (Figure 3.1(a)). For each URL discovered by the crawler, there is a corresponding node

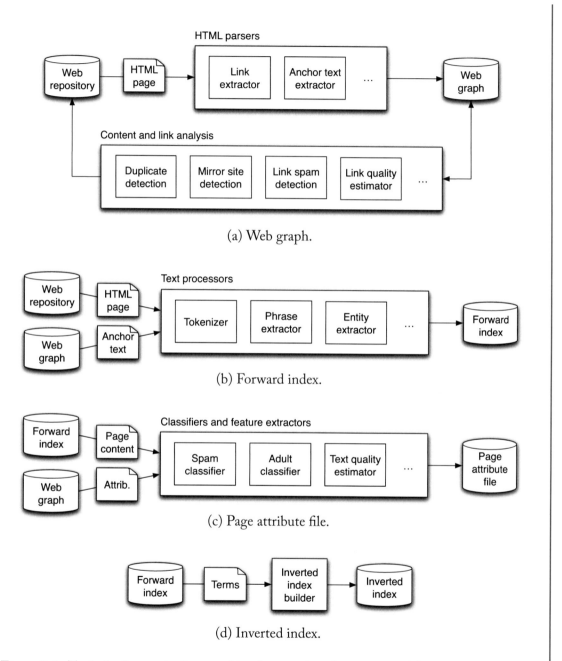

Figure 3.1: Typical software pipelines used in a large-scale indexing system. There are four main data structures created: web graph, forward index, page attribute file, and inverted index.

in the web graph. Every node maintains a number of attributes about its respective URL, such as the predicted geographical region and the importance of the URL (estimated through link analysis). The directed edges in the web graph are also associated with some data, such as the anchor text extracted from the respective hyperlinks. The web graph maintained by the search engine is only a partial representation of the actual, full graph of the Web. Therefore, some of the metrics computed using the web graph may not be exact (e.g., page importance measures and spam scores are computed based on partial information).

Certain graph measures need to be computed on coarse-grain web graphs. As an example, mirror sites can be detected using the site-level web graph, and the statistics about host responsiveness can be stored in the nodes of the host-level web graph. Therefore, several web graphs are built at different granularities (e.g., at the level of URLs, IP addresses, hosts, sites, and top-level domains). These web graphs may be exposed through a web service to facilitate access to this rich meta-data source from other systems. For example, using this service, the web crawling system can query the likelihood of a URL being spam or check whether a site is part of a link farm. In fact, some of the meta-data and scores obtained by processing the web graphs can be directly stored in the web repository (e.g., in the data structure that keeps the meta-data of web pages, as described in Section 2.4).

The page-level web graph can be quite large since it keeps information about every URL discovered by the crawler. Therefore, it has to be distributed over a large cluster of nodes. Accordingly, the algorithms that are used in processing the web graph need to be parallelized. The efficiency of these algorithms is important as their output is used to improve the quality of some time-sensitive tasks. For example, the page importance scores may be used for URL prioritization in the crawling system or as a ranking feature in the query processing system. Therefore, the algorithms are often substituted by faster approximations or alternatives that can attain slightly inferior yet reasonable quality.

As another source of information, a forward index is created, associating each web page with a list of terms contained in the page (Figure 3.1(b)). The forward index entry of a page is obtained by parsing the web page and extracting terms from its content. This typically involves the removal of HTML tags, parts of the boilerplate (e.g., the footer of the page), and non-static content (e.g., advertisements), leaving the main content of the page behind. In addition to the body, the terms can be extracted from the title, URL, and HTML header of the page. Besides single-word terms, the forward index may also contain phrases, named entities, and meta-words extracted from the page. Finally, the anchor text on the inbound links of the page is retrieved from the web graph, and the terms are extracted from the anchor text. Considering such anchor text as part of the referred web page makes sense as the anchor text often contains useful information about the page itself. Moreover, the anchor text can be used as surrogate content, enabling to search the pages in the frontier of the web crawler.

Given a web page, the terms can be extracted by tokenizing the textual content of the page (from all page sections mentioned above). White-spaces and most punctuation are removed as they are not considered as part of the terms. The punctuation is not removed if it is considered as part of the word (e.g., "e∗trade" or "c#"). The extracted terms are all converted to lower-case. In many research settings, the terms are also converted to their stems (e.g., the term "stemming" becomes "stem"). Stemming potentially increases the number of web pages that can match the query, but introduces some uncertainty into the matching process. In the context of web search, retrieving few, but highly relevant pages is more important than retrieving many pages with low relevance. Therefore, in web-scale indexing systems, stemming is usually not employed during the indexing process. Similarly, in most research settings, function words or stop-words (e.g., "the", "of", "and") are removed because these are usually uninformative terms when in isolation. In web-scale indexing systems, stop-words are indexed like any other term, as they may facilitate matching, especially in the case of queries that contain phrases (e.g., "the United States of America"). A related design choice needs to be made about whether the phrases (or word-based n-grams) in the page content should be treated as separate tokens and indexed individually. In general, including phrases in the index enables fast processing of phrase queries. However, since the index size increases considerably, the indexing of phrases is usually not preferred (with the exception of frequently queried phrases). Finally, depending on the language of the web page, language-specific processing rules or dictionaries may need to be used. For example, in some Asian languages, the tokenization procedure is different due to the absence of a word separator character. Moreover, stemming is language-dependent and, for many languages, it is more complex than in English.

A fundamental issue in indexing systems is the detection of duplicate pages. The web repository may contain multiple copies of a page because the same HTML content may be downloaded from different sources. Duplicate pages waste storage space and also incur unnecessary work for the indexing and query processing systems. A common technique used in duplicate page detection is to map the content of each web page to a hash value that represents it uniquely (with high likelihood). The uniqueness of a newly crawled web page can then be decided simply by checking if its hash value is equal to the hash value of any other page in the repository. This way, the duplicates can be grouped together and only a single copy of a page is kept in the repository. On the other hand, a large number of pages are actually not duplicates, but their content is only slightly different. Such pages are referred to as near duplicates, and their detection is a bit more involved. A commonly used technique for detecting near-duplicate pages is shingling. The n-shingling of a page is the set of all unique word-based n-grams (shingles) contained in the page. In a possible implementation of the shingling technique, a hash value is computed for every unique shingle of the page. These hash values are then sorted in increasing order, and the first few hash values at the head of the sorted list are selected. Two pages are assumed to be near duplicates if the Jaccard similarity of their selected hash values is above a threshold value. The threshold can be adjusted depending on the relative importance of false positives and false negatives. This process is illustrated with an example in Figure 3.2. Another solution to the near duplicate page detection problem relies on locality sensitive hashing, where content-wise similar pages are hashed into

p1: A B C D E F → 79 189 44 14 99 → H1 = {14, 44, 79} → J(H1,H2) = 1/5

p2: A B C X D Y F   79 189 84 68 6 87   H2 = {6, 68, 79}

Figure 3.2: A possible implementation for shingle-based near duplicate page detection. In this example, the shingles of a page are its word bigrams, and the similarity threshold is assumed to be 0.5. The two pages would not be detected as near duplicates since the Jaccard similarity between their smallest three hash values is below the threshold ($1/5 < 0.5$).

the same buckets with high probability. This way, the search space is limited to a small subset of pages.

A third data structure generated by the indexing system keeps various attributes of web pages as an array of records, one record per page (Figure 3.1(c)). This includes statistics, such as the number of characters and words existing in different sections of the page, as well as attributes computed by various classifiers, such as topic (e.g., sports, music, finance), language(s), genre (e.g., news, blog, commercial intent, porn). Moreover, a large number of features are extracted and stored. All of these features are query-independent features as they are extracted in an offline manner in the absence of a user query. They are later used by the query processing system, when ranking pages against queries. A small set of query-independent features are shown in Table 3.1. Some of these features (e.g., CTR values) are extracted by mining the query logs, which contain data about the users' past interactions with the search engine result pages. Query logs are usually stored in large data stores and processed in an offline manner. The extracted feature values are usually discretized into a smaller range so that they can be stored in fewer bytes. For example, log-transformed PageRank scores can be represented by integers from 0 to 255, thus requiring a single byte per score.

An important set of attributes involve the spam scores, which complement page importance scores. In general, the pages are classified, trying to detect four different types of web spam: content spam, link spam, cloaking/redirection spam, and click spam. In the case of content spam, the spam pages contain terms that are artificially placed in their content, as an attempt to increase their likelihood of matching search queries and obtaining a higher rank in search results. In the case of link spam, a target page is linked from many other pages, trying to increase the query-independent page importance scores computed by some link analysis metrics. In the case of cloaking, upon an HTTP request, the web server responds to the crawler with a normal looking web page, while it serves the users a spam page for the same URL request. In a similar technique, called redirection spam, the users are redirected to spam pages via JavaScript while the crawler receives the original page without being subject to redirection. Finally, click spam refers to fraudulent clicks on search results, aiming to boost the click popularity metrics computed for a web page or site.

Table 3.1: Example query-independent features that are stored in the page attribute file

Feature	Source	Description
Language	Page content	Language of the page
Length	Page content	Number of words or characters in the page
Content spam	Page content	Score indicating the likelihood that the page content is spam
Text quality	Page content	Score combining various text quality features (e.g., readability)
Link quality	Web graph	Page importance estimated based on page's link structure
CTR	Query logs	Click-through rate of the page in search results (if available)
Dwell time	Query logs	Average time spent by the users on the page
Page load time	Web server	Average time it takes to receive the page from the server
URL depth	URL	Number of slashes in the absolute path of the URL

The final and most important data structure is the inverted index (Figure 3.1(d)). Essentially, an inverted index is a mapping from each term in the web repository to the identifiers of web pages that contain the term. This data structure provides a very efficient mechanism to match the terms of a query against the terms in web pages. We use the toy example in Figure 3.3 to illustrate the process of converting a page collection to an inverted index. In this example, the collection and vocabulary sizes are $M = 6$ and $N = 14$, respectively. This may lead to the false impression that, in practice, the number of terms in the vocabulary is larger than the number of pages in the collection. Therefore, we should note that, in almost all practical scenarios, the collection size tends to be at least an order of magnitude larger than the vocabulary size.[1] Moreover, the inverted index shown in Figure 3.3 gives an extremely simplified view of what is actually created and deployed in commercial indexing systems.

3.2   INVERTED INDEX

Matching a query to a web page and computing the similarity between them are the two key operations in any query processing system. These operations form the main use case for the inverted index data structure. Therefore, before discussing the design of an inverted index, we will give some background on matching and similarity computations. Typically, a page is said to match a given query, if the page and the query have one or more terms in common. A quite naive way of matching a query against web pages is to scan the text of all pages in the repository and record the identifiers of pages whose terms overlap with the terms in the query. Obviously, this approach does not scale beyond very small page collections since it is too slow to scan large page collections entirely. As we will discuss below, an inverted index allows us to limit the search space to a much smaller subset of pages.

[1]According to the Heaps' law, the number of unique terms in a collection grows much slower than the number of pages in the collection.

Page 1: The smallest planet is Mercury.

Page 2: Jupiter, the largest planet!

Page 3: Neptune is smaller than Uranus,

    but is heavier.

Page 4: Uranus is smaller than Saturn.

Page 5: Neptune is heavier than Uranus.

Page 6: Jupiter > Saturn > Uranus > Neptune

(a) A six-page web collection.

Page 1: the smallest planet is mercury

Page 2: jupiter the largest planet

Page 3: neptune is smaller than uranus

    but is heavier

Page 4: uranus is smaller than saturn

Page 5: neptune is heavier than uranus

Page 6: jupiter saturn uranus neptune

(b) Normalized web pages.

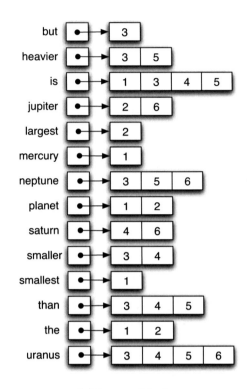

(c) Inverted index.

Figure 3.3: A sample web page collection and the corresponding inverted index.

Let us consider a page collection $C = \{p_1, p_2, \ldots, p_M\}$ with M pages and a vocabulary $V = \{t_1, t_2, \ldots, t_N\}$ of N terms that are extracted from the pages in C. As we mentioned before, an inverted index provides a mapping from each term $t \in V$ to a subset $\mathcal{L}(t) \quad C$ of pages. Given a query $q = \{q_1, q_2, \ldots, q_Q\}$ with Q terms, we can find the pages that contain at least one of the query terms by

$$\bigcup_{i=1}^{i=Q} \mathcal{L}(q_i), \tag{3.1}$$

or find the pages that contain all query terms by

$$\bigcap_{i=1}^{i=Q} \mathcal{L}(q_i). \tag{3.2}$$

These two operations are known as disjunctive and conjunctive matching, respectively. Both types of matching can be performed very efficiently using an inverted index.

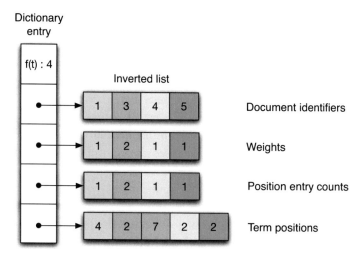

Figure 3.4: The dictionary entry for the term "is" and the corresponding inverted list (see Figure 3.3). This inverted index stores more information than the simpler inverted index in Figure 3.3. The four lists in the drawing are actually parts of a single, contiguous inverted list, but they were shown separately for the sake of presentation. Each color indicates the information related to a different page.

The similarity between a given query q and a matching page p can be computed in many different ways. A commonly used technique is to assign an offline-computed weight $w(t, p)$ to every term t in the page, indicating the term's importance. The similarity $s(q, p)$ between query q and some matching page p is then computed by summing these weights over all terms in the query:

$$s(q, p) = \sum_{i=1}^{i=Q} w(q_i, p). \tag{3.3}$$

There are many ways to compute the weight $w(t, p)$. Most techniques combine within-page term frequency values and collection-wide term frequency values. For example, in one variant of TF-IDF (term frequency-inverse document frequency) weighting, this weight is computed as

$$w(t, p) = \frac{f(t, p)}{|p|} \, \log \frac{M}{f(t)}, \tag{3.4}$$

where $f(t, p)$ denotes the number of times term t appears in p, $f(t)$ denotes the number of pages that contain t, and $|p|$ denotes the number of unique terms in p.

Armed with this information, we can now start providing more details about the inverted index data structure. An inverted index has two parts: inverted lists that keep certain statistics about the occurrence of the terms in the pages and a dictionary providing a mapping from the terms to the inverted lists (see Figure 3.4). For every term in the collection vocabulary, there is a corresponding inverted list. Similarly, the dictionary contains N entries (one entry per term).

The dictionary is used to access the inverted list of a given term. Each dictionary entry maintains the number of pages that contain its respective term (i.e., the $f(t)$ value), a pointer to the start of the inverted list, and potentially other meta-data about the term. The dictionary can be implemented as a hash table or B+ tree, keyed by the hash values of terms, as a trie data structure, or as a sorted array. The hash table and B+ tree implementations are usually more flexible and efficient. Since the dictionary is frequently accessed during the query processing to look up the terms, it is usually retained in the memory if possible.

The inverted list of a term has three sublists: document identifier, weight, and position lists. The document identifier list keeps the identifiers of pages in which the term occurs. The weight list keeps the term's weight in each of those pages. Finally, the position list keeps the positions at which the term appears in each page. A number of design choices need to be made to organize the information in these lists. These choices determine the compactness of the index as well as the time spent in matching and similarity computations. Therefore, the inverted index design has important implications for the scalability and efficiency of the query processing system.

The document identifiers are integer values uniquely selected from the $[1, M]$ range. An important design choice is about how the entries in the document identifier lists are ordered (the entries in other lists are sorted accordingly). One possibility is to sort the entries in increasing order of document identifiers. This kind of identifier-sorted lists enable the use of compression algorithms, leading to a relatively more compact index, yet allowing certain efficiency optimizations in query processing. Another possibility is to sort the entries in decreasing order of the associated term weights. This kind of weight-sorted lists enable the use of certain early termination heuristics in matching and similarity computations, eliminating the need to process all list entries. However, such lists are difficult to compress. A hybrid approach between identifier-sorted and weight-sorted lists is the impact-sorted lists. In this case, the term weights are quantized, and each page is assigned to an impact bucket according to its quantized term weight. The buckets in an inverted list are sorted in decreasing order of their impact. Within an impact bucket, the entries are sorted in increasing order of document identifiers. Impact-sorted indexes provide good compression rates as well as good query processing speed. Most indexing systems implement id-sorted or impact-sorted inverted lists. This choice is mainly due to the benefits of compression, which we will discuss in the next section.

Another design choice is about what kind of information should be stored in the weight lists. One alternative is to store the bare term frequencies (the $f(t, p)$ values). Since the term frequencies are usually very small integer values, they are highly amenable to compression. Another alternative is to store the length-normalized term frequencies (the $f(t, p)/|p|$ values). This eliminates the need to access the page attribute file to read the page length ($|p|$) at query processing time. A further alternative is to store the fully computed weights (the $w(t, p)$ values), turning the

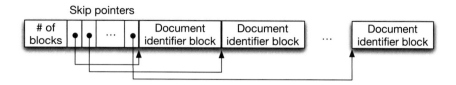

Figure 3.5: Skip pointers facilitate access to parts of an inverted list (assuming that the inverted list entries are sorted in increasing order of document identifiers).

online term weight computation (see Eq. (3.4)) into a simple lookup. In practice, most indexing systems prefer to store the term frequencies, delaying the computation of term weights to the query processing phase.

Finally, a position list keeps all occurrences of its respective term in individual pages. The positions at which the term appears in the term sequence of a page are represented by integer values starting from 1. For each page, there is a separate chunk of position information (e.g., in Figure 3.4, there are four pages and hence four chunks). Within a chunk, the positions are sorted in increasing order. The start address of each chunk can be maintained in an index. Alternatively, the length (in bytes) of each chunk can be recorded, and the start address of a chunk can be determined by accumulating these length values (Figure 3.4 assumes this approach). The positions of a term within a page may be further organized according to the sections of the page (e.g., URL, title, body, and anchor text). The position and section information are used during the query processing to match phrases against pages or to compute certain query-dependent features that exploit the proximity of the query terms in a page.

The document identifier, weight, and position lists are all organized under the assumption that the values will be accessed sequentially, starting with the first value in the list. If the pages are organized by increasing document identifier, a limited form of random access can be provided by splitting the inverted lists into smaller blocks and introducing pointers to the start address of these blocks. These so-called skip pointers allow jumping over the blocks and enable certain optimizations during the similarity computations. The use of skip pointers is illustrated in Figure 3.5.

Another useful optimization stems from the observation that the distribution of $f(t, p)$ values is very skewed. That is, there are many rare terms that occur in only a few documents, thus having very short inverted lists. The inverted list of each such term can be stored in a special format in the respective dictionary entry, overwriting the space allocated to list pointers. This reduces the size of the inverted index and allows fast access to the inverted lists of many terms.

The preferred method to store an inverted index on disk is to keep the entries of each inverted list in a contiguous byte array, enabling faster data transfer from the disk. As we will see in Section 3.5, however, in certain cases, some inverted lists may need to be spread over multiple

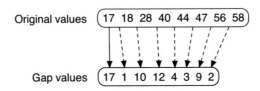

Figure 3.6: An example illustrating gap encoding. The first value is not encoded to facilitate decoding of gaps.

fragments and stored in a non-contiguous manner. A possible implementation in such cases is to use a linked list representation, where individual list fragments are linked to each other by disk address pointers.

3.3   COMPRESSING AN INVERTED INDEX

Compression is an important technique employed in indexing systems to reduce the size of an inverted index. A smaller inverted index implies that a larger portion of it can be kept in the memory, reducing the need to perform costly disk accesses. Even if the inverted lists have to be retrieved from the disk, the cost of reading the lists from the disk becomes much smaller. Obviously, once compressed, accessing the information encoded in the inverted index requires decompressing the inverted lists, partially or entirely. Therefore, compression leads to some computational overhead during the processing of queries. Overall, however, the efficiency gains in accessing the inverted lists justify the use of compression. In the context of inverted index compression, the performance of an algorithm is measured by three different metrics: compression rate, compression speed, and decompression speed.

In general, inverted lists are composed of highly compressible integer sequences. Certain compression algorithms operate on monotonically increasing integer sequences (e.g., Elias-Fano encoding and binary interpolative coding). Therefore, these algorithms are a good fit for compressing document identifier or position lists. The majority of the compression algorithms, however, deal with gap-encoded sequences, rather than the original integer sequences. Gaps refer to the differences between the consecutive values in a sorted list of integers (see Figure 3.6). In general, gap encoding increases the compressibility of inverted lists further as the obtained values are smaller integers which can be represented using fewer bits.

Once the gaps are computed for a given list, a compression algorithm can be applied to compress these gap values. In practice, there are a large number of inverted list compression algorithms with efficient implementations. The simplest compression algorithms are nonparametric (e.g., unary, Gamma, and Delta), while slightly more complex algorithms are parametric (e.g., Golomb encoding), i.e., they make use of the statistical properties of the list that is being com-

Table 3.2: Properties of some well-known compression algorithms

Encoding	Input sequence	Output	Parameters	Encoded values
Unary	gaps	bit-aligned	nonparametric	individual values
Gamma	gaps	bit-aligned	nonparametric	individual values
Delta	gaps	bit-aligned	nonparametric	individual values
Variable byte	gaps	byte-aligned	nonparametric	individual values
Golomb	gaps	bit-aligned	parametric	individual values
Simple-9	gaps	word-aligned	parametric	blocks of values
PForDelta	gaps	bit-aligned	parametric	blocks of values
Binary interpol.	monotonic sequences	bit-aligned	parametric	bisections
Elias-Fano	monotonic sequences	bit-aligned	parametric	entire sequence

pressed to improve the compression rate. While these algorithms encode individual gap values, more advanced algorithms compress larger blocks of gap values (e.g., PForDelta), providing better compression rates. Some algorithms produce byte- or word-aligned codes (e.g., variable byte encoding and Simple9), generally achieving competitive compression rates, yet with better decompression speed.

The performance of the compression algorithms shows great variation depending on the distribution of the values being compressed. Therefore, different algorithms may be preferred when compressing document identifier, weight, and position lists. In the case of term frequencies, since most term frequencies are already very small values, gap encoding is less useful and term identifiers can be compressed using simple encodings, such as unary encoding. Properties of some well-known compression algorithms are provided in Table 3.2.

The compressibility of document identifier sequences can be further improved by reassigning the document identifiers in a way that leads to smaller gaps than the original values. This technique is known as document identifier reassignment (see Figure 3.7 for an example). Besides increasing the compressibility of the index, document identifier reassignment may help to reduce the processing time of queries too. This is because, after reassignment of identifiers, pages that are relevant to the same queries tend to get together in consecutive blocks of an inverted list and hence most blocks can be skipped without decompressing them.

Fundamentally, the document identifier reassignment problem is similar to the sparse matrix compression problem. Since the problem is NP-hard, the existing solutions are all heuristics. A possible solution is to formulate the document identifier reassignment problem as a traveling salesman problem (TSP). This solution relies on creating a page similarity graph, where a weighted edge between two vertices indicate the similarity between the respective pages. The similarity between two pages can be computed in terms of the overlap between their terms. After the graph

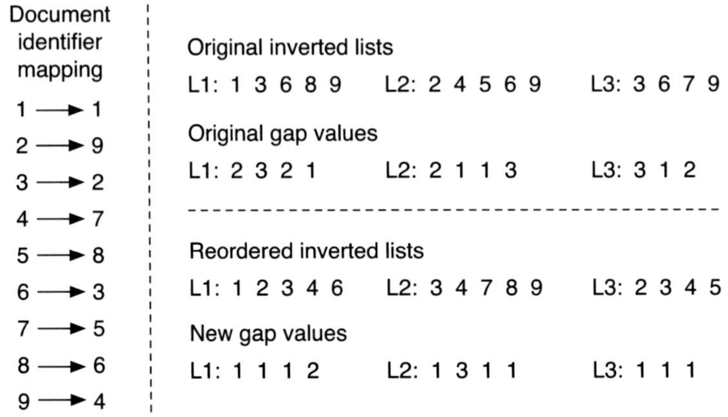

Document
identifier
mapping

Original inverted lists

L1: 1 3 6 8 9   L2: 2 4 5 6 9   L3: 3 6 7 9

1 ——► 1

2 ——► 9   Original gap values

3 ——► 2   L1: 2 3 2 1   L2: 2 1 1 3   L3: 3 1 2

4 ——► 7

5 ——► 8   Reordered inverted lists

6 ——► 3   L1: 1 2 3 4 6   L2: 3 4 7 8 9   L3: 2 3 4 5

7 ——► 5   New gap values

8 ——► 6   L1: 1 1 1 2   L2: 1 3 1 1   L3: 1 1 1

9 ——► 4

Figure 3.7: Reassigning document identifiers can lead to smaller gap values.

is created, a route is found by running the maximum TSP algorithm on the similarity graph. The document identifiers are then assigned to the pages according to their order in the obtained route.

Another possible solution relies on clustering. In this case, the pages are first clustered according to their content similarity. The clusters are placed in a sequence depending on the order they are created. Finally, consecutive identifiers are assigned to the pages that fall in the same cluster, starting with the first cluster in the sequence.

Both solutions mentioned above increase the likelihood of having consecutive document identifiers in the same inverted lists, thus increasing the number of small gaps and the compressibility of the inverted index. However, it is difficult to scale these techniques to work with a web-scale page collection. A relatively simpler, yet effective, solution is to order pages based on their URLs. In this case, the pages are first sorted according to the "reverse hostname:port/path" values. The document identifiers are then assigned sequentially starting from the first page in the sorted list. This technique exploits the fact that web pages with similar URLs tend to have similar content as well.

It is relatively uncommon to see document identifier reassignment techniques being implemented in indexing systems. This may be because the effect of document identifier reassignment on the overall index compression rate is limited, potentially because of the following two reasons. First, grouping pages with similar content may tend to place similar term weights or position sequences consecutively, increasing the compressibility of these sequences as well. However, the improvement is limited relative to that for document identifier sequences. Second, document identifier lists occupy a relatively small space in an inverted index compared to the position lists.

3.4   CONSTRUCTING AN INVERTED INDEX

In most commercial indexing systems, the inverted index is constructed and deployed periodically due to the practical issues involved in maintaining the index online (see Section 3.5). Queries continue to be evaluated over an older version of the index until a new index is deployed. It is important to reduce the duration of the index deployment cycle since a stale index may harm the search quality and hence the user experience. For example, the search results may contain deleted web pages or some recent web pages may be missing, even though those pages were downloaded by the crawling system.

Converting a page collection to an inverted index is similar to computing the transpose of a sparse matrix. For small collections, this is a rather straightforward task since the entire page collection can be read in a single pass to create an in-memory document-term matrix, which is then inverted in memory. For large page collections, this approach is clearly not feasible, and out-of-core index construction techniques are needed. The index construction techniques are distinguished depending on the number of passes made on the page collection, as illustrated in Figure 3.8.

In two-pass index construction, in the first pass, the pages are sequentially read from the disk (Figure 3.8(a)). For each term extracted from the content of pages, certain collections statistics are maintained. In general, it is sufficient to record the number of pages that contain the term and the number of times the term appears in each page. These statistics provide information about how much space the inverted list of each term will occupy on disk. Once all pages are read and the final term statistics are obtained, a skeleton of the inverted index is created on disk. For each term, an in-memory pointer is maintained to the start address of its inverted list template on disk. In the second pass, the pages are read from the disk one more time. For each term encountered in a page, the document identifier, weight, and position information is written to the corresponding inverted list on disk. The information collected for each term can be buffered and written later in larger chunks to prevent excessive disk accesses.

In one-pass index construction, a partial inverted index is built in the memory while pages are read from the disk (Figure 3.8(b)). Every time the memory space is exhausted, the partial index built so far is flushed to the disk, and a new partial index starts to grow from scratch. Once the final page is read and processed, the current partial index is written to the disk. This process creates multiple partial inverted indexes on disk. In the final step, these partial indexes are merged into a full index, which is again written to the disk. One-pass index construction is usually faster than two-phase index construction. Two-pass index construction may perform better in systems with very limited memory. In such systems, one-phase index construction leads to too many partial indexes on disk, increasing the cost of merging.

In large-scale indexing systems, the index is built and deployed on thousands of nodes. Therefore, besides being scalable and efficient, the indexing system needs to be tolerant to hardware, software, and network failures. The Hadoop technology, which provides a distributed storage and processing platform for very large data sets, is commonly used to create an inverted index

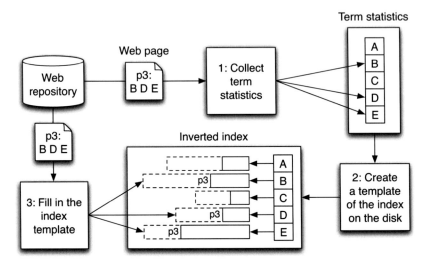

(a) Two-pass index construction.

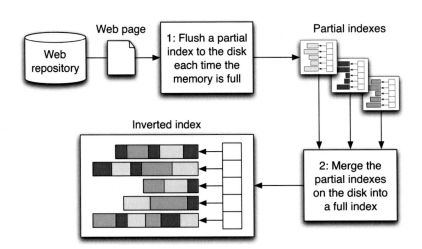

(b) One-pass index construction.

Figure 3.8: Alternative techniques to construct an inverted index.

in a scalable, fast, and reliable manner.[2] On this platform, converting a page collection into an inverted index is a surprisingly simple task. Figure 3.9 illustrates the basic process: the mappers read the web pages and emit (term, document identifier) pairs, where the keys are the terms and the

---

[2]More information about Hadoop is available at https://hadoop.apache.org.

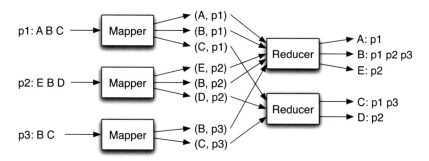

Figure 3.9: Inverted index construction using the Hadoop platform. Three web pages (p1, p2, p3) are converted into the inverted lists corresponding to five different terms (A, B, C, D, E). The example assumes that the terms are mapped to reducers in a random manner.

values are the document identifiers. The emitted pairs are partitioned across the reducers according to their terms. The document identifiers associated with each term are grouped at different reducers. Finally, the reducers output the list of document identifiers for each term, one term at a time.

3.5   UPDATING AN INVERTED INDEX

Periodically rebuilding the index guarantees a certain level of freshness. For the main web search index, a deployment cycle of a few hours may be considered reasonable. However, in time-sensitive search verticals (e.g., news search or tweet search), the freshness requirements are even stricter. That is, potential modifications in the page repository (i.e., page insertions, deletions, and updates) should be reflected to the inverted index as quickly as possible.

Incorporating term-related information obtained from newly downloaded web pages into an existing inverted index is not a trivial task. This is because the existing inverted lists are likely to be tightly packed on disk and there is no space to accommodate new entries inside the lists. In practice, there are several solutions to overcome this issue. All of these solutions rely on growing a so-called delta index in the memory. A delta index is a relatively small inverted index that keeps the information about the most recently added web pages. Once the delta index is large enough it is written to the disk, and a new delta index starts to be built. The available solutions differ in the way this delta index is merged with the main inverted index on the disk. Several index maintenance approaches are illustrated in Figure 3.10.

A simple approach is to store each delta index as a separate index on the disk, besides the main index (Figure 3.10(a)). This approach has low index maintenance overhead since all needs to be done is to swap the delta index from the memory to the disk. However, this solution has a major

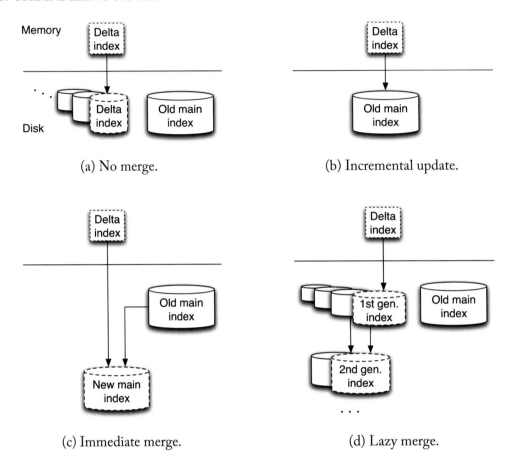

(a) No merge.

(b) Incremental update.

(c) Immediate merge.

(d) Lazy merge.

Figure 3.10: Different approaches for maintaining an inverted index as new pages are inserted.

drawback. As the number of delta indexes on disk increases, the inverted lists are fragmented over too many small indexes. This, in turn, increases the cost of query processing since multiple disk accesses are required for each query term when evaluating a query. Therefore, this solution is not scalable in the long term. Eventually, the entire index has to be built from scratch.

A different approach is to update the main index *in situ* (Figure 3.10(b)). This approach requires allocating some buffer space at the end of every inverted list when the main index is created. During the merge, the entries in the delta index are appended to these buffer spaces and are processed as additional entries in the matching and similarity computations. This kind of incremental updates eliminate the need to store one or more delta indexes separately. Moreover, since the updates require accessing only specific parts of the main index, the need to read/write the entire main index at each merge is avoided. However, this approach comes with two drawbacks. First, the updates are performed on inverted lists, which are concurrently accessed by the query processing system, i.e., the two systems compete for the same resources. Therefore, the index updates may have a negative effect on the query processing time of individual queries. Second, certain inverted lists eventually run out of free buffer space, and new space needs to be allocated for such lists. A possible strategy here is to maintain the lists in a contiguous manner at all times. This requires moving the data around and efficient space management. An alternative is to spread the lists over multiple fragments. Although this is more flexible and avoids data movements, it degrades the query processing performance since a larger number of disk seeks are needed to read fragmented inverted lists.

Another approach is to merge the delta index and the main index in memory and write an up-to-date main index to the disk (Figure 3.10(c)). In this case, only a single index is maintained on disk at all times. Therefore, the query processing cost is not affected. However, every merge operation requires reading and writing the entire index. A variation of this approach is to maintain multiple generations of delta indexes on disk (Figure 3.10(d)). In this case, the delta indexes are lazily merged over time to create larger indexes. This variation creates a trade-off between the index maintenance overhead and query processing cost.

Besides handling page insertions, a proper index maintenance technique should also handle page deletions. A naive option here is to maintain the identifiers of deleted pages in an in-memory data structure and filter them out from the ranked results during the query processing step. A better approach is to introduce some sort of garbage collection mechanism and regularly remove these pages from both in-memory and on-disk indexes (e.g., when the fraction of deleted pages is above a threshold). This kind of garbage collection can be performed on-the-fly during the regular merge operations as well.

3.6   PARTITIONING AN INVERTED INDEX

In large-scale web search engines, the inverted index is partitioned on a search cluster involving thousands of nodes. As we will see in Section 4.4, a query is evaluated by processing the partial indexes on the nodes concurrently and merging the results obtained on each node. The way the index is partitioned on the nodes has significant implications for the performance of the query processing system. In general, there are two ways of partitioning an inverted index: document-based or term-based partitioning.

In document-based index partitioning, the page repository is partitioned among the nodes such that all information related to a page is stored on the same node. Therefore, all inverted list entries belonging to a page become available in one node. A document-partitioned index can be formed by assigning a disjoint subsets of pages to the nodes and then building a local inverted index at each node using the pages assigned to the node. The pages can be distributed on the nodes in a round-robin fashion, according to their identifiers. This usually leads to good load balance in terms of index sizes.

In term-based index partitioning, the vocabulary is partitioned over the nodes such that the inverted list of each term is assigned to a particular node as a whole. A term-partitioned index can be obtained by building a global index on the entire page collection and then assigning subsets of terms and the corresponding inverted lists in the global index to the nodes. Since the inverted list sizes are highly skewed, random or round-robin assignment of terms may lead to high storage load imbalance on the nodes. Therefore, assignment techniques that explicitly try to reduce the load imbalance are preferred (e.g., bin packing).

In the case of document-based partitioning, reading the complete inverted list of a term requires performing multiple disk accesses since the entries of an inverted list are scattered on many nodes. In the case of term-based index partitioning, an inverted list can be accessed with potentially fewer disk seeks since all entries in the list are stored on the same node. In either case, the amount of data read from the disk is the same. However, the disk I/O is likely to be faster in the case of document-based partitioning, since the inverted lists are shorter and they can be accessed concurrently. In terms of space consumption, both techniques have additional requirements. In the case of document-based partitioning, the vocabulary needs to be replicated on all nodes. In the case of term-based partitioning, the page attribute file is replicated on all nodes. Assuming that the pages are already uniformly distributed on the nodes, document-based partitioning offers some simplicity in index construction as the local indexes can be built on each node independently. Similarly, maintaining the local indexes is easier since all index updates related to a page are performed in the same node. Constructing a term-partitioned index, on the other hand, requires coordination and communication among the nodes.

Document- and term-based index partitioning strategies are illustrated in Figure 3.11, continuing the example in Figure 3.3. Different aspects of the two techniques are compared in Table 3.3. We will later provide a separate comparison in the context of query processing since that comparison requires the knowledge of parallel query processing, which will be introduced in Section 4.3. For the same reason, we survey the related work on index partitioning strategies in Section 4.7, instead of in Section 3.7.

3.7   LITERATURE ON INDEXING

There are several publications that provide an architectural overview of indexing systems used for commercial or academic purposes. The details of the indexing system underlying IBM's intranet search engine, Trevi, were disclosed in Fontoura et al. [2004]. Besides index construction, this paper provides some clues about the document processing pipeline used in Trevi as well. The Maguro indexing system, which is part of the Bing serving stack, was briefly described in Risvik et al. [2013]. This system is specifically designed to index long tail content, and it differs from most other commercial indexing systems in that it opts for term-based index partitioning and has a strong preference toward indexing phrases.

Data structures play a key role in indexing systems. Signature files [Faloutsos, 1992] and suffix trees/arrays [Gonnet et al., 1992, Manber and Myers, 1990] were old alternatives to inverted indexes [Harman et al., 1992]. Given the size of page collections that are indexed by web search engines today, the inverted index stands as a much more scalable data structure [Zobel et al.,

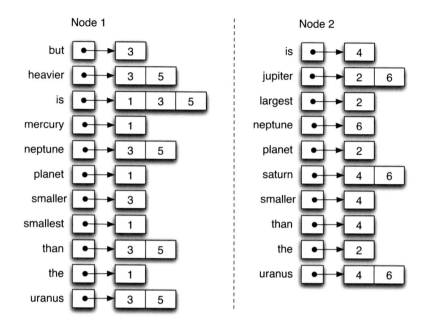

(a) Document-based partitioning (the document identifiers are assigned to nodes in a round-robin fashion).

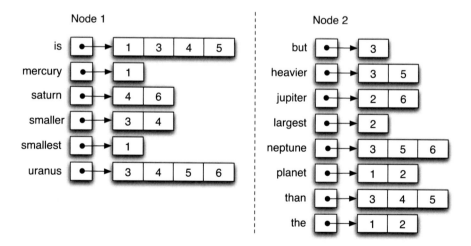

(b) Term-based partitioning (the terms are assigned to nodes according to their first letters, which are mapped to nodes in a round-robin fashion).

Figure 3.11: Two different index partitioning strategies are illustrated using the example inverted index in Figure 3.3.

Table 3.3: Comparison of index partitioning strategies (K: number of nodes in the search cluster, M: number of pages in the collection, N: number of terms in the vocabulary)

Aspect	Index partitioning strategies	
	Document-based	Term-based
Vocabulary	Replicated on each node – $O(KN)$	Partitioned over the nodes – $O(N)$
Page attribute file	Partitioned over the nodes – $O(M)$	Replicated on each node – $O(KM)$
List access	Many concurrent partial list accesses – $O(K)$	Single list access – $O(1)$
List I/O from the disk	Shorter lists – takes less time to read	Longer lists – takes more time to read
Index construction	Simpler – local indexes	More complex – global index
Index maintenance	Updates on a single node – $O(1)$	Updates on many nodes – $O(K)$

1998]. Although the inverted index is considered as a standard data structure in web search, it is still open to improvements. A relatively old improvement was introducing skip pointers into inverted lists [Moffat and Zobel, 1996]. A more recent technique that exploits the query distribution to optimize the placement of skip pointers can be found in Chierichetti et al. [2008]. More recently, the treap data structure, which was investigated in Konow et al. [2013], was shown to achieve promising results in terms of space and time efficiency. In Hsu and Ottaviano [2013], different variants of suffix trees were used for efficient top-k query completion.

There is a vast literature on near duplicate page detection. Shingles were used for near-duplicate page detection in Broder et al. [1997]. An alternative technique that relies on locality-sensitive hashing was described in Charikar [2002]. A large-scale evaluation of these two techniques is available in Henzinger [2006]. Other near duplicate page detection techniques can be found in Chowdhury et al. [2002], Manku et al. [2007], Sood and Loguinov [2011], and Theobald et al. [2008].

The details of commercial spam classifiers are usually kept confidential as spammers and web search engines are in a constant arms race. Nevertheless, there is a lot of academic work [Spirin and Han, 2012] on web spam, especially about content and link spam detection. An early work on detecting content spam is Fetterly et al. [2004a], which applies statistical analysis on page properties, such as linkage structure, page content, and page evolution. In Ntoulas et al. [2006], a handful of other features were used in a content spam classifier. In [Urvoy et al., 2008], a radically different approach was taken in that the spam pages are detected based on the similarity of their HTML styles, instead of textual content. The technique proposed in Najork [2012] was specifically designed to detect web pages whose content is created by stitching together pieces of text taken from other web pages. A well-known technique to detect link spam relies on trust propagation over the web graph starting from a small seed set of trustable pages. The quality of this algorithm, known as TrustRank [Gyöngyi et al., 2004], was improved in two consecutive works [Krishnan and Raj, 2006, Wu et al., 2006]. In Anti-TrustRank [Krishnan and Raj, 2006], the propagation was performed on the inverted web graph, starting from a set of known spam pages. In topical TrustRank [Wu et al., 2006], the seed set was partitioned according to the topic of pages and the trust scores were calculated for each topic separately. In a different line, link-based features were extracted to learn spam classifiers [Becchetti et al., 2008]. The work in Cheng et al. [2011] suggested detecting link spam by mining suspicious links posted on web forums. Techniques for detecting cloaking were investigated in Wu and Davison [2006] and Lin [2009]. The components of a bot designed for click fraud attack were described in Daswani and Stoppelman [2007]. A general web spam classifier was built in Liu et al. [2012] using features related to the page access patterns of users.

The older compression algorithms in the literature usually operate on individual gap values. Among these, unary, gamma [Elias, 1975], and delta [Elias, 1975] encodings are non-parametric while Golomb [Golomb, 1966] and Rice [Rice and Plaunt, 1971] encodings are parametric. More recent algorithms compress the values in longer sequences or blocks (e.g., PForDelta [Zukowski et al., 2006], OptPFD [Yan et al., 2009b], Simple-9 [Anh and Moffat, 2005], Simple-16 [Zhang et al., 2008], and VSEncoding [Silvestri and Venturini, 2010]). Some algorithms produce byte-aligned codes (e.g., variable byte encoding [Cutting and Pedersen, 1990] and its variants [Stepanov et al., 2011]). A number of algorithms operate on the original monotone integer sequences, instead of gaps (e.g., binary interpolative encoding [Moffat and Stuiver, 2000]). Along this line, the recent research on inverted list compression, based on the Elias-Fano representation [Elias, 1974, Fano, 1971], provides promising results [Ottaviano and Venturini, 2014, Vigna, 2013]. Compression algorithms for encoding position lists were proposed in Yan et al. [2009a]. In Ottaviano et al. [2015], the blocks were compressed by different encoders that are selected according to the distribution of terms in queries.

The document identifier reassignment problem was formulated as a TSP problem in Shieh et al. [2003] for the first time. Due to the high time complexity of this solution, the follow-up work focused on improving its scalability. In Blanco and Barreiro [2006], the singular value decomposition was applied to reduce the dimensionality of the space in which pages are represented. In Ding et al. [2010], locality sensitive hashing was used to find the most similar pages in an efficient manner. Both optimizations led to faster similarity computations and a more sparse similarity graph on which the TSP algorithm can be executed more efficiently. Clustering-based solutions were proposed in Blandford and Blelloch [2002] and Silvestri et al. [2004]. The former work used top-down hierarchical clustering, while the latter work investigated both top-down and bottom-up clustering algorithms. A simple solution, based on ordering pages according to their URLs, was proposed in Silvestri [2007]. Two different works investigated the interplay between document identifier reassignment and gap compression techniques [Arroyuelo et al., 2013, Yan et al., 2009b]. The technique proposed in Arroyuelo et al. [2013] created long gap runs of all 1s in frequently accessed inverted lists, and run-length encoding was used to compress these runs. In Yan et al. [2009b], the existing techniques for compressing document identifiers and term frequencies were optimized for the case where the gaps are obtained after document identifier reassignment.

Index construction approaches can be classified, depending on the number of passes on the page collection, as two-pass approaches [Fox and Lee, 1991, Moffat and Bell, 1995] or one-pass approaches [Harman and Candela, 1990, Heinz and Zobel, 2003, Moffat and Bell, 1995]. A two-pass, in-memory inverted index construction approach was proposed in Fox and Lee [1991]. In this approach, after completing a first pass on the page collection, an in-memory bit vector was allocated to store the list entries obtained in the second pass. This approach was extended in Moffat and Bell [1995] such that the preallocated vector was stored on disk to provide scalability with increasing collection size. A one-pass index construction approach, which maintains the list entries in an on-disk linked list, was described in Harman and Candela [1990]. This approach, however, is somewhat inefficient due to the large number of random disk accesses required. More efficient, one-pass approaches were described in Heinz and Zobel [2003], Melnik et al. [2001], and Moffat and Bell [1995]. The approach in Moffat and Bell [1995] proposed flushing sorted runs of list entries to the disk and then merging these runs by k-way merge to obtain the complete inverted lists. While the approach in Heinz and Zobel [2003] was also based on k-way merging of partial indexes flushed to the disk, it did not require keeping the vocabulary in the memory since the vocabulary was flushed as part of the partial indexes. A similar two-phase approach can be found in Melnik et al. [2001]. This approach increases the efficiency of partial index creation by improving the concurrency of three consecutive tasks (reading web pages, processing their content, and flushing the created partial index). Finally, the scalability of building an inverted index on the Hadoop platform was investigated in Mccreadie et al. [2012].

Given a page collection that is initially distributed on the nodes of a parallel system, the work in Ribeiro-Neto et al. [1999, 1998] presented various techniques to create a term-partitioned inverted index on the same parallel system. The technique in Ribeiro-Neto et al. [1998] assumed that the index fits in the main memory. Three alternative techniques were described in Ribeiro-Neto et al. [1999] assuming that the partial indexes are maintained on disk before they are merged. The techniques differed in their selection of where the memory buffers of list entries are maintained and where the partial indexes are stored on disk (in the local or remote nodes).

The immediate merge strategy was first described in Clarke et al. [1994]. The lazy versions of this strategy, which maintains multiple partial indexes on disk and delay the merge operation, were proposed much later under the names of geometric partitioning [Lester et al., 2008] and logarithmic merge [Büttcher and Clarke, 2005]. The incremental update strategy was studied in Brown et al. [1994] and Shieh and Chung [2005], tackling the problem of estimating how much free space should be allocated at the end of each list. Comparisons between different index maintenance strategies indicated the superiority of the immediate merge strategy over the incremental update strategy [Lester et al., 2004, 2006]. However, more recent experiments demonstrated that the incremental update strategy has become more feasible due to the advance of flash SSDs [Jung et al., 2015]. A large number of studies employed hybrid approaches that apply the incremental update strategy or a merge-based strategy to inverted lists, selectively, depending on criteria such as term access frequency [Gurajada and P, 2009] or the inverted list size [Büttcher and Clarke, 2008, Büttcher et al., 2006, Cutting and Pedersen, 1990, Margaritis and Anastasiadis, 2009, Tomasic et al., 1994]. Two studies proposed flushing the inverted lists in the delta index selectively, rather than flushing the entire delta index [Büttcher and Clarke, 2008, Margaritis and Anastasiadis, 2009]. The index maintenance strategies proposed in Büttcher and Clarke [2005] and Guo et al. [2007] supported both page insertions and deletions.

Although it is an old book, most of the indexing algorithms and data structures described in Witten et al. [1999] are still used in commercial indexing systems. A more recent reference with a good coverage of indexing techniques is Zobel and Moffat [2006]. A summary of the indexing literature is provided in Table 3.4.

3.8   OPEN ISSUES IN INDEXING

A potential venue for research is to design strategies for optimal placement of inverted lists on disk. In most work in the literature, the inverted lists are stored on disk as contiguous byte sequences, ignoring the block structure of modern disks. The co-access patterns of inverted lists can be used to pack concurrently accessed lists in the same disk blocks. This may provide gains in terms of transfer times while the inverted lists are read from the disk during the query processing.

Inverted index maintenance is a well-studied topic. However, almost all works so far has dealt with the insertion operations, i.e., the scenarios where the information about newly acquired pages are incorporated into an existing inverted index. In dynamic environments like the Web, handling page deletion and modifications is also important, and a full-fledge indexing system should support all kinds of index update operations. The design of efficient data structures that can support page deletions, insertions, and modifications at the same time is an open research issue.

Another open research issue is the engineering of features that accurately measure the users' interest in a web page and the extraction of those features in real time. Such features are especially important for quick promotion/demotion of viral or ephemeral web pages in search rankings. In practice, these features can be obtained from social media (e.g., Twitter) as well as the physical network (e.g., deep packet inspection on the routers). In any case, the time-sensitive nature of these features require devising web-scale, streaming data processing techniques to efficiently compute and maintain the feature values over time.

Although index construction has been considered in the case of parallel architectures (e.g., a cluster of nodes), it is not yet obvious how an index can be constructed, distributed, and maintained over multiple geographically distant data centers. In this scenario, the high network cost involved in moving the data between different data centers necessitates devising strategies for efficient coupling of the indexing system with the crawling and query processing systems.

Table 3.4: A summary of the literature on indexing

Topic	Reference	Short description
General overview	[Witten et al., 1999]	Book on indexing data structures
	[Zobel and Moffat, 2006]	Survey on inverted indexes
Indexing system design	[Fontoura et al., 2004]	Overview of the indexing system in Trevi
	[Risvik et al., 2013]	Overview of the indexing system in Maguro
Indexing data structures	[Faloutsos, 1992]	Signature files
	[Gonnet et al., 1992]	Suffix trees and arrays
	[Harman et al., 1992]	Inverted indexes
	[Zobel et al., 1998]	Compares inverted indexes and signature files
	[Moffat and Zobel, 1996]	Skip pointers (placed based on the inverted list length)
	[Chierichetti et al., 2008]	Skip pointers (placed based on the query distribution)
	[Konow et al., 2013]	Treap data structure
	[Hsu and Ottaviano, 2013]	Various trie data structures for top-k query completion
Near duplicate detection	[Broder et al., 1997]	Makes use of shingles
	[Charikar, 2002]	Makes use of locality-sensitive hashing
	[Henzinger, 2006]	Comparison of Broder et al. [1997] and Charikar [2002]
	[Chowdhury et al., 2002]	Makes use of collection statistics
	[Manku et al., 2007]	Uses the simhash fingerprinting technique
	[Sood and Loguinov, 2011]	Improves [Manku et al., 2007]
	[Theobald et al., 2008]	Semantic preselection of shingles
Web spam detection	[Spirin and Han, 2012]	Survey of web spam detection techniques
	[Fetterly et al., 2004a]	Content spam
	[Ntoulas et al., 2006]	Content spam (learns a classifier using link features)
	[Urvoy et al., 2008]	Content spam (uses the HTML style)
	[Najork, 2012]	Content spam (uses shingles)
	[Gyöngyi et al., 2004]	Link spam (TrustRank)
	[Krishnan and Raj, 2006]	Link spam (Anti-TrustRank)
	[Wu et al., 2006]	Link spam (topical TrustRank)
	[Becchetti et al., 2008]	Link spam (learns a classifier)
	[Cheng et al., 2011]	Link spam (mines the links on web forums)
	[Wu and Davison, 2006]	Cloaking detection
	[Lin, 2009]	Cloaking detection
	[Daswani and Stoppelman, 2007]	Details of a click spam bot
	[Liu et al., 2012]	General web spam (uses page access patterns of users)
Inverted index compression	[Elias, 1975]	Gamma and delta encoding
	[Golomb, 1966]	Golomb encoding
	[Rice and Plaunt, 1971]	Rice encoding
	[Zukowski et al., 2006]	PForDelta encoding
	[Yan et al., 2009b]	OptPFD encoding
	[Anh and Moffat, 2005]	Simple-9 encoding
	[Zhang et al., 2008]	Simple-16 encoding
	[Silvestri and Venturini, 2010]	VSEncoding
	[Cutting and Pedersen, 1990]	Variable byte encoding
	[Stepanov et al., 2011]	SIMD-optimized variants of variable byte encoding
	[Moffat and Stuiver, 2000]	Binary interpolative encoding
	[Elias, 1974]	Elias-Fano encoding

	[Fano, 1971]	Elias-Fano encoding
	[Ottaviano and Venturini, 2014]	Partitioned Elias-Fano indexes
	[Vigna, 2013]	Quasi-succinct indexes
	[Yan et al., 2009a]	Algorithms for compressing position lists
	[Ottaviano et al., 2015]	Chooses among multiple encoders
Document identifier reassignment	[Shieh et al., 2003]	The first TSP-based solution for the problem
	[Blanco and Barreiro, 2006]	Efficiency optimizations before TSP
	[Ding et al., 2010]	Efficiency optimizations before TSP
	[Blandford and Blelloch, 2002]	Top-down clustering solution
	[Silvestri et al., 2004]	Top-down and bottom-up clustering solutions
	[Silvestri, 2007]	Proposes a simple solution based on URL ordering
	[Arroyuelo et al., 2013]	Document identifier reassignment and compression
	[Yan et al., 2009b]	Document identifier reassignment and compression
Inverted index construction	[Fox and Lee, 1991]	One-pass index construction (in-memory)
	[Moffat and Bell, 1995]	One-pass and two-pass index construction (on-disk)
	[Harman and Candela, 1990]	Two-pass index construction (on-disk)
	[Heinz and Zobel, 2003]	Two-pass index construction (on-disk)
	[Melnik et al., 2001]	Two-pass index construction (on-disk)
	[Mccreadie et al., 2012]	One-pass index construction (Hadoop)
	[Ribeiro-Neto et al., 1998]	Parallel term-partitioned index construction (in-memory)
	[Ribeiro-Neto et al., 1999]	Parallel term-partitioned index construction (on-disk)
Inverted index maintenance	[Clarke et al., 1994]	Immediate merge
	[Lester et al., 2008]	Lazy merge (geometric partitioning)
	[Büttcher and Clarke, 2005]	Lazy merge (logarithmic merge), support for deletions
	[Brown et al., 1994]	Incremental update
	[Shieh and Chung, 2005]	Incremental update
	[Lester et al., 2004]	Immediate merge, no merge, incremental update
	[Lester et al., 2006]	Immediate merge and incremental update
	[Jung et al., 2015]	Incremental update (on flash SSDs)
	[Gurajada and P, 2009]	Hybrid (according to term frequency)
	[Büttcher and Clarke, 2008]	Hybrid (according to list size), selective flushing
	[Büttcher et al., 2006]	Hybrid (divide the list into a long part and a short part)
	[Cutting and Pedersen, 1990]	Hybrid (uses B-tree and heap file)
	[Margaritis and Anastasiadis, 2009]	Hybrid (according to list size), selective flushing
	[Tomasic et al., 1994]	Hybrid (disk space management policies for long lists)
	[Guo et al., 2007]	Lazy merge, support for deletion operations

CHAPTER 4

The Query Processing System

The query processing system acts as the front door of a search engine. For each user query submitted through some search interface, this system performs three fundamental tasks. First, the intent of the received query is interpreted and the query is rewritten in some internal representation, potentially after some normalization and enhancement. Second, the rewritten query is evaluated by processing the inverted index and other data structures in the backend search system, retrieving a set of best-matching result pages that are ranked in decreasing order of their estimated relevance to the query. Third, some information is compiled about the retrieved pages (e.g., title, URL, and summary) and this information is presented on the SERP to the user who issued the query.

As we briefly mentioned in Section 1.1, the success of a search engine is highly related to its query processing system's ability to find pages that well match the information needs expressed in user queries and rank those pages in a meaningful way. Despite the immense size of the Web, for most queries, there are only few pages that are highly relevant (this often leads to an analogy between web search and searching for a needle in a haystack). Moreover, as some leading search engines continue to increase their search quality standards, the users become more reluctant to scan beyond the few top results on the SERPs. Therefore, besides correctly identifying the information needs of the users, the query processing system must be highly effective in matching the pages in its index to the identified information needs. In this context, the effectiveness can capture other aspects of search quality, such as the diversity of the presented results or the descriptiveness of result summaries, not solely the result relevance, although we often use this term to imply relevance.

In addition to being effective, a query processing system has to be efficient. The efficiency of a query processing system is usually measured by two complementary metrics. The first metric is the response latency, which is a measure of the speed at which the query processor retrieves and serves the matching results to the users. Reducing the response latency is important because, either consciously or subconsciously, the users' engagement with the search engine can be negatively affected due to high response latency. The second metric is the query processing throughput, which is a measure of the number of queries that can be answered by the query processing system within a given time period (a common unit for this metric is queries per second). An efficient query processor should be able to continue to operate under heavy query traffic, sustaining a certain peak throughput.

In practice, there is often an interplay between the efficiency and effectiveness requirements. For example, a sudden peak in the query traffic may increase the average response latency of the query processing system. As another example, the processing of a query may be terminated early, potentially degrading the quality of retrieved results, if the response latency is expected to be too high for the query. In general, the trade-off between these metrics leads to interesting alternatives in the design of a query processing system. Therefore, most optimizations are devised keeping potential trade-offs in mind. In this chapter, we review some architectural and algorithmic optimizations employed in large-scale query processing systems, with special emphasis on the efficiency.

The rest of the chapter is organized as follows. In Section 4.1, we describe the architecture of a typical query processing system. Sections 4.2 and 4.3 present the key design issues in sequential and parallel query processing, respectively. Some high-level architectural optimizations are presented in Section 4.4. In Section 4.5, we discuss caching, which is a key factor for reducing the query workload of the backend search systems. Multisite distributed query processing is covered in Section 4.6. Section 4.7 reviews the literature on query processing systems. The chapter ends in Section 4.8 with some open research problems.

4.1 BASIC QUERY PROCESSING ARCHITECTURE

In accordance with the three tasks mentioned in the previous section, we will investigate the query processing system in three parts: query interpretation, result retrieval, and result preparation systems, each of which may be replicated on a number of nodes or clusters. The query interpretation and result preparation systems can be considered as part of the search frontend, while the result retrieval system forms the search backend. A high-level picture of the interaction among these three systems is given in Figure 4.1.

The query interpretation system is responsible for transforming the original user query into an internal representation specified by the search engine. The input to this system is the original user query expressed by keywords and the output is a rewritten query expressed in an internal query language. The query interpretation system is implemented as a pipeline of small software modules, resembling the document processing pipelines we described earlier in Section 3.1. The tasks per-

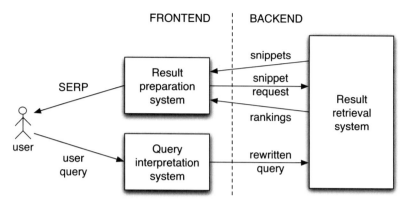

Figure 4.1: The three subsystems in a query processing system and their interaction with each other.

formed along this pipeline include normalization, spelling correction, segmentation, stemming, annotation, and term expansion, among others. In practice, it is vital to align the functionality provided in the query interpretation and document processing pipelines. That is, the same kind of manipulation should be performed on web pages and queries to prevent potential inconsistencies that may arise while matching pages to query terms. For example, if the stop words are eliminated while indexing the web pages, they should be removed from the queries as well. Otherwise, queries involving stop words may not match any result page if they are evaluated in conjunctive mode.

The diagram in Figure 4.2 illustrates the text processing modules often used in a query interpretation system. Some of these modules perform simple text processing tasks (e.g., normalization and segmentation), while others perform slightly more involved NLP tasks (e.g., spelling correction and stemming) or semantic analysis (e.g., term expansion using ontologies). As illustrated in the figure, some of the modules in the pipeline may decide to alter or not alter the input query (or do both). For example, the spelling correction module may pass as input to the segmentation module both the normalized query and a spelling-corrected version of it. As a result, multiple variants of the original user query are created and carried along the pipeline. The query rewriting module in the last step combines these variants under a single, internal query. As an example, the user query "amusement arcades in New York" may be converted to the internal query "AND(OR(PHRASE(amusement arcade), PHRASE(video arcade)), LOCATION(new york))" after several modifications. In this example, four modifications were performed on the original query: (i) the stop word "in" was removed, (ii) the term "arcades" was converted into its singular form "arcade," (iii) "amusement arcade" was detected as a phrase and expanded to "video arcade," and (iv) "New York" was detected as a location and converted to lowercase. Once a query is rewritten in such a form it is passed to the result retrieval system.

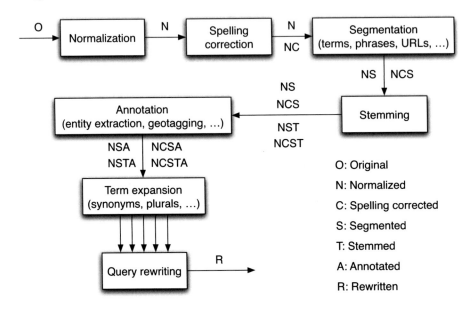

Figure 4.2: Possible software modules in a query interpretation system.

The result retrieval system is responsible for evaluating the rewritten query on one of the main web search clusters as well as a number of vertical search clusters. Every search cluster computes a ranking of web pages after evaluating the rewritten query by processing the index and returns a ranking to the result preparation system. The result retrieval system also hosts a serving system that serves the textual content of web pages. Among all the tasks in the query processing system, generating the result ranking on the main web search cluster is the most time-consuming task. Fortunately, there is a wide range of potential efficiency optimizations that can speed up evaluation of queries. We will defer the discussion of the result retrieval system to the succeeding sections. The interaction between the search clusters in the result retrieval system and the frontend systems is depicted in Figure 4.3.

The result preparation system takes the rankings retrieved from different search clusters and prepares the final SERP returned to the user. This system performs mainly three tasks. First, since multiple rankings are retrieved from different search clusters, the result preparation system acts as an aggregator. It makes decisions about which rankings (that is, the search modules we discussed in Section 1.1) should be displayed on the SERP and where they should be slotted. Usually, this kind of optimization is performed through online machine learning models that are optimized based on the user interaction with the individual search modules on the SERP. Second, the result preparation system issues additional requests to the backend search system to gather the textual data associated with the pages. This is because the results retrieved from the

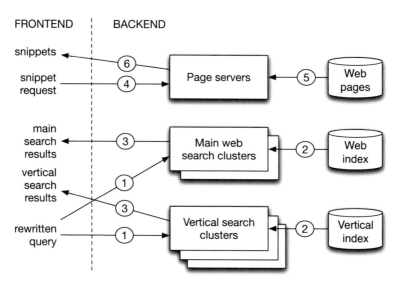

Figure 4.3: A more detailed view of the way the frontend systems interact with the backend result retrieval system.

ranking system include only the document identifiers and some additional data (e.g., scores), but not the text of pages. The retrieved textual information includes the title and URL of pages as well as a short summary of each page, referred to as snippet or caption. Since the snippets of a page are generated in a query-specific manner (the snippets include parts of the text that contain the query terms), they are generated online by the document servers in the backend system. In practice, generating the snippets of many result pages can be time consuming. However, since the pages are likely to be distributed on different document servers, the snippet generation can be performed concurrently for different pages. Finally, further filtering or reranking may be performed on the retrieved results. For example, the results may uniqued such that no more than a few results are displayed from the same host (typically, at most two results are shown per host). The system may go one step further and apply a result diversification algorithm on the final ranking. In result diversification, the rank of a page may be demoted if there is a content-wise similar page that is higher ranked than the page. This kind of diversification is useful especially when the query intent is ambiguous, and hence it can be applied selectively only on such queries. Applying a diversification algorithm on a large result set may be expensive. Therefore, these algorithms are usually applied on the top-ranked pages (e.g., top 100) as a post-processing step.

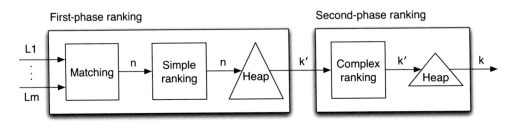

Figure 4.4: Two-phase ranking within a search node. The first-phase ranking system restricts the initial set of n matching pages, which are obtained by processing the inverted lists $(L1, \ldots, Lm)$ associated with the query, to k' candidate results (k' is typically around hundred results per search node). The second-phase ranking system determines the final top k results returned to the user (k is typically ten per search node or is slightly larger).

4.2   QUERY PROCESSING ON A SEARCH NODE

The objective in query processing is to estimate the relevance of a set of pages to a given user query, rank the pages in decreasing order of their relevance, and present a small set of most relevant pages to the user. In practice, simple scoring schemes (e.g., the one described in Section 3.2) are not sufficient to obtain high-quality rankings, especially when today's web search standards are considered. Therefore, search engines use complex ranking models generated by machine learning systems to rank the pages. While processing a query, the ranking model is used to generate a relevance score between the query and different web pages. In theory, all pages in the web index can be evaluated against the query using the learned ranking model. However, in practice, it is not possible to score every page this way since the execution time would be prohibitively high. Therefore, most search engines adopt a two-phase ranking framework. In the first phase of this framework, a simple yet sufficiently accurate scoring technique is used to select a small subset of potentially relevant pages from the entire page collection. By filtering potentially less relevant pages, this phase helps to reduce the computational cost of the second phase. In the second phase, the pages selected in the first phase are reranked by a complex but much more accurate ranking model. The final ranking is obtained by sorting the page scores computed in the second phase in decreasing order. This two-phase ranking architecture is illustrated in Figure 4.4, which shows how the number of matching result pages is gradually reduced.

The first phase involves matching the web pages to the query and computing a relevance score for each matched page (the two tasks are usually interleaved). The pages can be matched to a query by iterating over the inverted lists corresponding to the query terms and keeping track of the document identifiers encountered in the list entries. The matching can be performed in conjunctive mode or disjunctive mode, and the list entries may be sorted in different ways, as we discussed before in Section 3.2. Moreover, during the query processing, the inverted lists

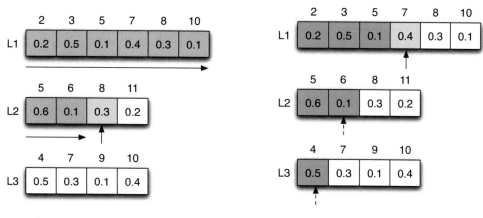

a) Term-at-a-time traversal.   b) Document-at-a-time traversal.

Figure 4.5: Two different ways of traversing inverted lists. The examples show a snapshot of the progress of score computation. The numbers outside the lists are document identifiers while the numbers inside are the scores. In the term-at-a-time traversal example, lists L1, L2, an L3, were so far fully processed, partially processed, and not processed, respectively. In the document-at-a-time traversal example, all document identifiers less than or equal to 6 were already processed, and the currently processed document identifier is 7.

can be traversed one document-at-a-time or one term-at-a-time, as illustrated in Figure 4.5. In document-at-a-time traversal, the lists are traversed simultaneously and each page's score is fully computed before considering the next page. In this approach, it is sufficient to maintain only the top k scores during the score computations. In term-at-a-time traversal, only one inverted list is actively processed at a given time. This approach requires maintaining the score of each page in an accumulator data structure. In general, all of these alternatives create an interesting design space for efficiency optimizations. In fact, another possibility is score-at-a-time processing, where the list entries are processed in decreasing order of their score contributions. This is feasible only for weight-sorted or impact-sorted lists.

Assuming that matching is done in disjunctive mode, a very naive implementation for scoring computations relies on using an array of accumulators. In this implementation, the score contributions coming to each page from different list entries are accumulated in the array entries. After all inverted lists associated with the query are processed and the final page scores are obtained, the accumulator array is traversed to find the highest-scoring pages and sort them. The problem with this implementation is the use of a very large accumulator array, which needs to be stored in the memory and traversed for every query. The implementation can be improved by accumulating the scores in a hash table and extracting only the top k entries (instead of sorting the entire array).

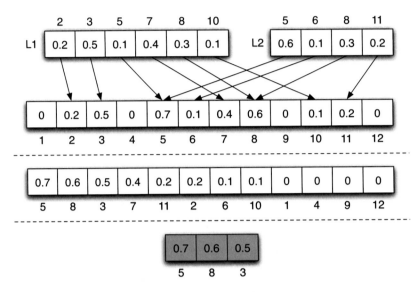

(a) Accumulator array implementation.

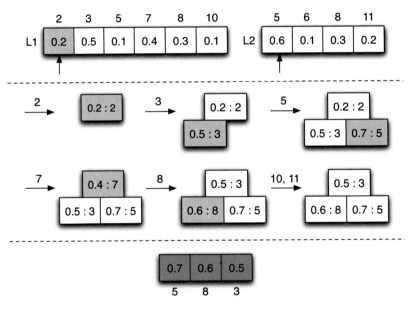

(b) Min-heap implementation.

Figure 4.6: Possible implementations of scoring computations in the first-phase ranking. The example assumes that $k = 3$ pages are retrieved as the top results.

A much more efficient implementation (assuming document-id-sorted lists and document-at-a-time traversal) relies on using a min-heap priority queue that maintains the scores of a small number of top-scoring pages at any time (at most k scores). The heap data structure guarantees that, once all list entries are processed, those that remain in the heap are the highest-scoring pages. These two implementations are illustrated in Figure 4.6.

More efficient implementations make use of certain properties of inverted lists to avoid redundant score computations. Common optimizations include skipping score computations for relatively unimportant pages, early terminating score computations under certain effectiveness guarantees, or bounding the number of allocated accumulators at runtime. An optimization is said to be set-safe if it returns the right top k, but without a guarantee on the right score ordering. It is said to be rank-safe if the score ordering is also guaranteed, even if the exact values of the scores have not been fully computed. Two example query processing optimizations are illustrated in Figure 4.7.

In most commercial query processing systems, the queries are matched in conjunctive mode (by default), which tends to emphasize precision, because this mode of processing leads to better search result quality than the disjunctive mode, which tends to emphasize recall. The inverted lists are traversed using the document-at-a-time traversal strategy, rather than term-at-a-time traversal because the former strategy is amenable to better efficiency optimizations. The first-phase score of a page is usually computed as a linear combination of its query-dependent and query-independent scores. This way, both the relevance of the page to the query and its query-independent importance or quality are captured. Incorporation of query-independent scores increases the quality of pages that are passed as input to the second-phase ranking. However, in this case, the scoring optimizations become much less effective as the score bounds are usually not tight enough.

The learning models used in the second-phase ranking are trained by trying to minimize a function indicating the difference between the relevance scores generated by the learned model and some ground-truth data indicating the relevance between a sample set of web pages and queries. The ground-truth data can be obtained editorially or through explicit user feedback on SERP (e.g., clicks or dwell time). The training models use a large number of features obtained from the query, pages, as well as other sources. As we discussed in Section 3.1, the query-independent features are extracted by the indexing system in an offline manner. The query-dependent features are computed before evaluating the query by the machine-learned ranking model. Important query-dependent features include various query-page similarity scores (e.g., BM25, which is a popular scoring function for relevance ranking) and some term proximity scores that measure the proximity of the query terms in the page content. The query-dependent features are computed, separately, for different sections of the page (e.g., title, body, and anchor text).

The most commonly used learners in search engines include tree ensembles and neural networks. In practice, there may be more than one learning model trained and deployed in production. For example, depending on device type (e.g., PC, tablet, or mobile), different learning

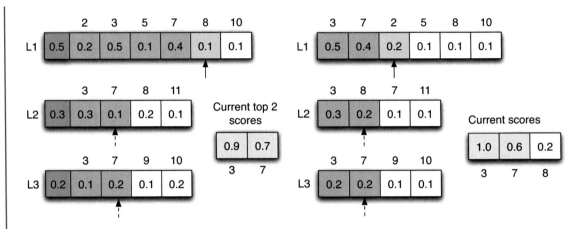

(a) A dynamic index pruning approach.　　　　(b) An early termination approach.

Figure 4.7: Two possible query processing optimizations. In (a), the maximum score contribution of each list is statically stored in the purple entries. At this state, the current page in L1 (the document identifier is 8) can be skipped without computing its complete score since it is guaranteed that the maximum score the page can attain $(0.1 + 0.3 + 0.2)$ will be less than the bottom score (0.7) currently stored in the heap (assuming we are retrieving the top $k = 2$ pages). In (b), the traversal of the lists is terminated early since it is guaranteed that no new page can make it into the final ranking. In this case, the lists are sorted in decreasing order of their score contributions and a separate accumulator is maintained for each page seen so far. Although the set of top-k pages are correctly computed, the scores may be incomplete and hence the final ranking may not be rank-safe.

models can be trained, exploiting historical usage data specific to a particular device type. Queries submitted from a particular device type are evaluated using the learning model trained for that device type.

Although the first-phase ranking limits the candidate set of pages to be ranked by the second-phase, the efficiency of the second-phase ranking is still important. For efficient second-phase ranking, two types of approaches may be followed. The first type includes offline approaches that aim to create compact learning models that reduce the overhead of score computations in the second-phase ranking without sacrificing much from the result quality. The second type includes online approaches. For example, the score computations may be terminated early for a selected subset of pages, which are not likely to make it to the final top-k ranking. In this case, the termination decisions can be made by heuristics that exploit simple features, such as the current score of pages.

Certain result retrieval systems may provide support for result customization, including regionalization or personalization. Regionalization is a form of coarse-grain result customization, where the ranking is altered to retrieve web pages relevant to the users in a particular geographical region (e.g., a Mexican user may be returned relatively more pages written in Spanish or hosted on servers located in Mexico). The regionalization of results can be achieved simply by boosting the first-phase scores of certain pages via heuristics that exploit the matching between the region/language of the user and the region/language of the page. Alternatively, the region and language information can be used as features in the second-phase ranking. Personalization is a form of fine-grain result customization, where further features are incorporated into the ranking such that a specific result ranking is generated for each user. The additional features may be obtained from the demographics of the user, self-declared interests, and the query and click history associated with the user. The benefits of personalization are debatable. Since the results tend to be biased toward what the user has already found (due to the use of user's search history), personalization may decrease the chance of accessing new information. Moreover, as a result of personalization, the number of unique result rankings increases. This, in turn, reduces the effectiveness of result caches, which are very important for decreasing the query workload of the backend search systems (Section 4.5).

4.3   QUERY PROCESSING IN A SEARCH CLUSTER

One of the efficiency objectives in search engines is to maintain the response latency of the query processing system at reasonable levels, as the web index increases in size. This objective is achieved by distributing the inverted index over the nodes of a search cluster and evaluating each query in parallel by processing this distributed index. A search cluster, depending on the size of the web index, may include hundreds or thousands of computers. As the web index grows in size, the search cluster is grown proportionally by adding new nodes such that the index size on each node remains roughly the same. Another efficiency objective is to scale the query processing throughput with increasing query traffic volume. This is achieved by replicating the web index on multiple search clusters and directing a portion of the query traffic to each search cluster.

The queries can be scheduled on search clusters in different ways. A simple solution is to assign queries to search clusters in a round-robin fashion. A better solution takes into account the current workload of search clusters. That is, newly arriving queries are scheduled on the least loaded clusters. More complicated scheduling approaches may even shuffle the order of queries in the waiting queue to prioritize certain queries (e.g., those that are likely to suffer from high response latency). Yet another approach is to map each query to a unique search cluster, based on the hash of the query string. In practice, this approach is both simple and effective. It provides fairly good load balance while increasing the utilization of the inverted list caches in the cluster nodes.

The queries are processed in parallel on the nodes of a search cluster. Each cluster has a special node, called the broker. The broker node receives queries from the query interpretation system in the frontend and communicates them to the search nodes in the cluster. It is also responsible for aggregating the partial rankings retrieved from each node into a final ranking and communicating this ranking to the result preparation system. In practice, a cluster can have more than one broker node. In an extreme case, each node in the cluster can act as a broker while also functioning as a query processor.

The inverted index can be distributed on the search nodes using either a document-based or term-based partitioning strategy, as we discussed in Section 3.6. The way the queries are evaluated on a search cluster depends on the adopted partitioning strategy. Two possible parallel query processing strategies are illustrated in Figure 4.8.

In the case of document-based partitioning (Figure 4.8(a)), the query is communicated to all search nodes. The first and second scoring phases are tightly coupled within each node. Every page matching the query goes through first-phase scoring. Next, the second-phase scores are computed for a small set of selected pages which have relatively high first-phase scores. Each node then returns its top-k ranking (document identifiers and the associated second-phase scores) to the broker. The top-k ranking for the entire search cluster is obtained at the broker by sorting the pages in decreasing order of their second-phase scores.

In the case of term-based partitioning (Figure 4.8(b)), the broker locally maintains some information about the mapping of terms to search nodes. Moreover, the document identifier space is partitioned among the search nodes evenly and the page attribute file is stored accordingly, such that each node has sufficient information to compute the second-phase score of the pages assigned to itself. Queries are evaluated in two consecutive steps. In the first step, the broker communicates the query to the search nodes whose inverted index may have a page matching the query. For every matching page encountered in a search node, a partial first-phase score is computed. The nodes return these partial scores to the broker, which aggregates them and obtains a global ranking of pages according to their complete first-phase scores. In the second step, the broker selects a number of top-ranked pages from the obtained ranking and communicates each selected page to the search node responsible for computing the second-phase score of the page. Every contacted node computes its local top-k ranking involving the second-phase scores. These rankings are then returned to the broker, which merges them into a top-k ranking for the entire search cluster.

In general, document-based partitioning provides intra-query parallelism as the same query can be processed on all nodes, at the same time. Term-based partitioning, on the other hand, provides inter-query parallelism since multiple queries can be concurrently processed in the search cluster, mostly in the absence of intra-query parallelism. Term-based partitioning works well in distributed search systems connected with a fast network while the speed of disk seeks is more critical in the case of document-based partitioning. A comparison of the two strategies is given in Table 4.1.

In most web search engines, document-based index partitioning is preferred due to several reasons. First, document-based partitioning leads to better computational load balance. Second, the brokers tend to be a bottleneck in the case of term-based partitioning. Third, document-based partitioning provides better scalability in terms of increasing number of nodes and collection sizes. Finally, document-based partitioning provides better fault tolerance than term-based index partitioning since the result quality is less likely to be affected in the case of node failures.

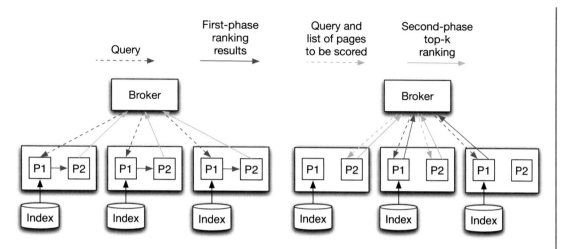

(a) Document-based index partitioning.   (b) Term-based index partitioning.

Figure 4.8: Parallel query processing under document- and term-based index partitioning. (a) The query is issued to all three search nodes and each node returns its second-phase ranking to the broker. (b) The query is issued to only the second and third nodes, which compute and return their first-phase rankings to the broker. The broker then initiates the second-phase ranking on the first and second nodes, which return their second-phase rankings to the broker.

Table 4.1: Comparison of parallel query processing strategies

Aspect	Parallel query processing strategies	
	Document-partitioned index	Term-partitioned index
Preferred hardware	Better with fast disks	Better with fast networks
Parallelism in query processing	Intra-query parallelism	Inter-query parallelism
Computational load imbalance	Lower	Higher
Potential for bottleneck at the broker	Less likely	More likely
Scalability with increasing query volume	Better	Worse
Effect of node failures on search quality	Lower	Higher

The trade-off here is between missing a few relevant documents (in the case of document-based partitioning) and missing an entire inverted list (in the case of term-based partitioning) when processing queries.

4.4   ARCHITECTURAL OPTIMIZATIONS

In certain search architectures, the search space is reduced by evaluating queries on smaller portions of the web index. The main objective in this kind of architectural optimizations is to reduce

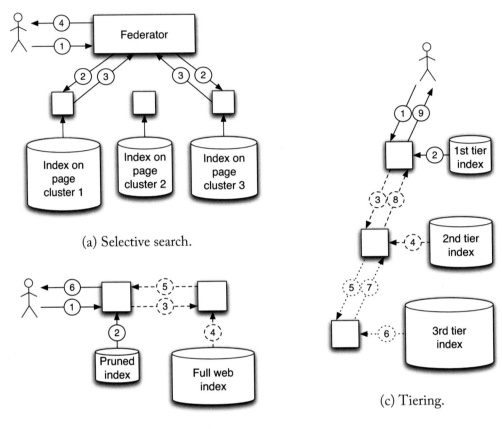

(a) Selective search.

(c) Tiering.

(b) Static index pruning.

Figure 4.9: Three different architectural optimizations for efficient query processing. The solid-line circles indicate the steps that are always executed by the technique, while the optional steps are indicated by the dashed-line circles.

the query workload of the result retrieval system as well as the query processing time, without hurting the quality of the retrieved search results. Depending on how the web index is organized, these optimizations can be classified under three headings: selective search, static index pruning, and tiering, which are illustrated in Figure 4.9 and discussed below.

The selective search architecture involves an offline clustering phase and an online retrieval phase. In the clustering phase, the pages in the web repository are grouped under a number of subcollections. The clustering can be based on the topical similarity of the pages or their co-occurrence in result rankings. An inverted index is then built on each subcollection, and each index is distributed on a number of search nodes. In the retrieval phase, a query is evaluated by processing a selected subset of these indexes and aggregating the retrieved results in a federating node. A subcollection and its respective index are selected only if the subcollection is estimated to contain some pages relevant to the query. A possible collection selection technique is to store the statistical summary of the terms in each collection on the federating node and estimate the relevance of subcollections by the similarity between these summaries and the terms in the query. Another option is to build separate, small inverted indexes on sets of pages sampled from each collection and process these indexes to find out how likely the collections are to yield pages matching the query. The example in Figure 4.9(a) illustrates a selective search architecture with three page clusters, only two of which are selected when processing the current query. This kind of selective search reduces the processing load since not all indexes are processed for every query. On the other hand, the search quality may be inferior compared to the one that can be obtained by processing the entire web index since some relevant results may be missed. Moreover, since the topic distribution of pages and queries can have a large variation, the processing load across the indexes may be highly unbalanced, and this may lead to an increase in the response latency of individual queries.

The idea in static index pruning is to construct an inverted index that keeps much less information than the full web index, and yet is almost as effective in answering queries. A pruned index can be obtained in various ways. The standard index pruning techniques rely on removing the least important inverted list entries from the full web index. In this architecture, queries are first evaluated by processing the pruned index. Then, a decision is made about whether the results obtained from the pruned index are good enough or some important results are missing. The full web index is processed only if the results obtained from the pruned index are found unsatisfactory. The success of static index pruning depends on the fraction of queries that can be answered solely by the pruned index. In theory, this architecture is expected to reduce the processing load as well as the average query processing time. However, if too many queries hit the full web index, the pruned index becomes an overhead and one may end up with an inefficient system. Some static index pruning techniques provide guarantees on the quality of the results obtained by processing the pruned index. However, most static index pruning techniques do not provide such a guarantee on search quality.

In tiering, the pages are partitioned into disjoint sets known as tiers, according to their importance, and an index is built on each set of pages. The importance of a page is usually determined by the likelihood of a page appearing in the search results or a link-based importance score. The early tiers are smaller in size and keep relatively more important pages, while the later tiers are gradually larger in size and keep less important pages. A query is processed by hitting the tiers in decreasing order of page importance and merging the results obtained from each tier on the way. After obtaining the results from a tier, a decision is made about whether the next tier should be hit or not. This fall-through decision can be made depending on the estimated search quality or simpler features, such as the total number of pages obtained so far from the former tiers, and has an impact on the search quality as well as performance. In tiering, there is no need to maintain a full web index, unlike static index pruning.

In large-scale query processing systems, tiering is usually preferred over selective search and static index pruning. This is mainly because tiering is easy to design and implement, as page importance is a readily available feature. Moreover, the skewed nature of page importance distribution helps the scalability of this architecture: unimportant pages are accessed with a relatively much lower likelihood. In the case of selective search, maintaining the load balance is difficult due to the changes in the query stream, while static index pruning requires using additional hardware and is relatively difficult to implement in practice.

4.5   CACHING

Caching is one of the oldest optimizations in computer systems. The basic idea behind caching is to store some data, selectively, in a small but fast storage system. The requests for data are processed by looking up the data first in this small storage system, called the cache. If the requested data is found in the cache, it can be readily served by the cache. Otherwise, the data is served by a larger storage system, which is located on a slower storage device. The former case is called a cache hit, while the latter case is called a cache miss. Cache storage systems are often organized in a hierarchy. Well known examples of cache hierarchies include those in operating systems (registers, L1 cache, L2 cache, memory, disk, and tape) and the Web (browser cache, web proxy, web server cache, and backend datastore).

Search engines make heavy use of caching. An important type of cache is the result cache, which is usually located in the frontend system or distributed over the broker machines. This cache keeps the results of previously processed queries. Upon a cache hit, the results of repeating queries are returned by the result cache, avoiding expensive query processing computations at the backend search system. Optionally, a result cache can be coupled with a score cache located at the search nodes. This cache may keep the precomputed relevance scores for the local top k results of queries to prevent redundant computation of relevance scores.

Another type of cache is the inverted list cache. This cache exists in the search nodes and keeps selected inverted lists in the memory. This way, during the query processing, the disk accesses are avoided for the selected lists. A possibility is to couple this cache with an intersection cache, which stores the precomputed intersections of frequently co-accessed lists. This cache speeds up query processing since fewer lists need to be traversed and the intersected lists are shorter.

Finally, selected web pages are cached. This is an in-memory cache located within the page storage system. It enables faster snippet generation, as the text of the pages do not need to be retrieved from the disk. We illustrate all types of caches mentioned above in Figure 4.10.

Caching becomes more beneficial especially when the data access has a heavily skewed distribution. This is exactly the case in search engines. Few queries are submitted many times, while many queries appear only a few times. For example, the most frequently issued query usually constitutes a surprisingly large fraction of the query volume, while queries that do not repeat

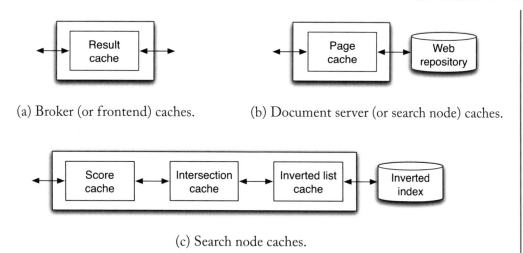

(a) Broker (or frontend) caches.   (b) Document server (or search node) caches.

(c) Search node caches.

Figure 4.10: Various types of caches found in a web search engine.

within a year after they are issued constitute almost about half of the volume. Although less skewed, similar observations can be made for inverted list and page accesses.

An important design issue in caching is to decide what parts of the data (or precomputed results) should be kept in the cache. In general, there are two features that are important in this decision: the frequency and recency of access to the data items (query results, inverted lists, or the textual content of pages). That is, data items that are more frequently or more recently accessed are usually preferred over other data items for caching. Based on these two features, we can classify the search engine caches into two as static and dynamic caches. The static caches keep frequently accessed data. These caches are created in an offline manner using historical access frequency information of data items. A static cache is filled with the most frequently accessed items, in a greedy fashion, until the cache capacity is met (if the item sizes are not equal, a better strategy is to fill the cache with items having the largest frequency/size ratio). Dynamic caches, on the other hand, try to capture the recency of data access and store more recently accessed items. In this respect, their maintenance is online. Items that are least likely to be requested in the near future are evicted from the cache to make space for other items. Since they capture different characteristics of the request stream, static and dynamic caching can be coupled within a hybrid caching framework. For large enough caches (e.g., on-disk caches), however, static caching may be redundant since a large dynamic cache is likely to contain most of the frequently accessed items anyway.

Result caches are on-disk, dynamic caches that can store the results of hundreds of millions of queries. Because of the high capacity of such caches, a large fraction of cached query results are never evicted from the cache. The web index, however, is continuously updated as new pages are inserted or deleted. During the course of time, the modifications in the index gradually increase the staleness of the results served by cache. A common technique to alleviate this staleness issue is to associate each cache entry with a time-to-live value, indicating when the entry should be considered as expired. A cache hit on an expired entry is treated like a cache miss and the cached query results are refreshed, i.e., the query is processed at the backend result retrieval system and the cached results are replaced with the newly obtained results. The results of selected queries may be refreshed even before they are expired, making use of the low-traffic times of the search backend. In this case, to prevent redundant query processing, the results of a query are refreshed only if they are estimated to be stale. An alternative technique to refreshing is to invalidate the cached results based on the feedback provided by the indexing system. For example, when a new page is inserted in the index, the indexing system notifies the result cache about which cache entries may have turned stale and must be invalidated. Although this kind of cache invalidation is potentially more accurate than selectively refreshing the query results, it is harder to implement since it requires interaction and coordination between two decoupled systems.

4.6   QUERY PROCESSING ON MULTIPLE SEARCH SITES

Search engines operate on massive data centers. In the basic case, all operations of the search engine may be centralized on a single site. That is, crawling, indexing, and query processing systems are all located on a single data center. In practice, search engines distribute their operations on multiple data centers that are potentially distant to each other. This kind of multisite search architectures offer increased fault tolerance and better business continuity. For example, when a data center becomes inaccessible due to a natural disaster, network partition, or some other reason, the user queries may still be served by the remaining data centers. Given multiple search sites, there are two alternative ways to process queries, depending on whether the index is replicated or partitioned across the data centers, as illustrated in Figure 4.11.

In the case of replication, the entire web index is made available on all sites by periodically reconstructing the index in one of the data centers and copying it to all sites. The queries are processed as in the case of a centralized architecture with a minor difference. Each search site processes queries issued from the region in which it is located. This reduces the average network latency between the sites and the users, thus reducing the average query response latency. If a data center is overloaded with too many queries, its query workload can be temporarily shifted to other data centers.

In the case of partitioning, the index is divided into multiple subindexes depending on the relevance of pages to a geographical region. That is, each search site maintains the pages relevant to its region and builds an index on those pages. This architecture is motivated by the fact that a large fraction of queries have local search intent and there is no need to search the entire index for relevant pages (nevertheless, a small portion of globally popular pages can be replicated on all sites and indexed separately). In this architecture, the processing of queries is slightly more complex. As in the case of the replicated architecture, the user space is partitioned among the search sites according to the geographical distance. Each search site acts as a local site for queries

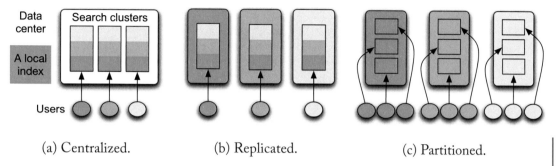

(a) Centralized.   (b) Replicated.   (c) Partitioned.

Figure 4.11: Query processing architectures assuming different data center and index configurations. There are nine search nodes in each configuration. (a) There is a single data center with three search clusters, each made up of three nodes. The index is fully replicated on the clusters and partitioned among the nodes. Queries are processed in the same data center, but potentially in different search clusters. (b) There are three separate data centers, each containing a single search cluster. The data centers serve queries originating from their region. (c) There are three separate data centers, each containing a single-node search cluster. Every data center maintains its own local inverted index, which is replicated on the search clusters. Since the local indexes are smaller, more queries can be evaluated concurrently.

originating from its region. A query is first issued to its local search site, which produces some results by processing its local index. Next, depending on the estimated quality of the retrieved results, a decision is made about whether the query should be forwarded to some of the remote search sites. If the query is forwarded to a remote search site, it is evaluated by processing the respective index. The results retrieved from the remote site(s) are merged with the local results and returned to the user. The query forwarding process is illustrated in Figure 4.12.

The key to the success of this architecture is to accurately estimate which search sites can contribute good-quality results to the final top-k result set of a query. Accurate estimations prevent redundant forwarding of queries to remote search sites while also avoiding degradation in search quality due to not forwarding. In this respect, the described architecture is similar to the selective search architecture presented in Section 4.4. If no remote sites are contacted during the process, the response latency of the query is reduced. In general, since the web index is only partially processed, the overall query workload is expected to be lower than that of the centralized or replicated architectures. Therefore, although this architecture does not provide any guarantee on the search quality, it can considerably improve the query processing efficiency.

Certain optimizations are possible when making the forwarding decisions. A potential optimization relies on storing on all search sites the maximum possible score contributions the indexed terms can receive from each local index. Assuming an additive scoring function like the one in Eq. 3.3, it now becomes possible to compute certain score upper bounds for a given query

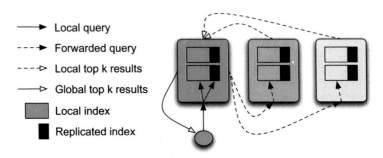

Figure 4.12: Query forwarding in a multisite web search engine with a partitioned index. A small fraction of highly popular pages are replicated and indexed on all search sites. In the example, the user query is issued to a local data center, which evaluates the query by processing its local and replicated indexes. The query is then forwarded to two remote data centers, which return their local top k results (the query is evaluated by processing only the local indexes). Finally, after merging all retrieved results with its local ranking, the initial data center returns a global top-k ranking to the user.

and guide the query forwarding decision according to comparisons between the computed score bounds and the scores obtained at the local site. For example, a query may be forwarded to a remote site only if the score upper bound associated with the remote site is larger than the kth score obtained at the local site.

4.7   LITERATURE ON QUERY PROCESSING

Google's search architecture was described first in Brin and Page [1998] and later in Barroso et al. [2003], the former focusing more on the software platforms and the latter focusing more on the hardware infrastructure. The query processing system of Maguro was described in Risvik et al. [2013]. In this system, each search node had a three-layer ranking stack (matching, simple ranking, and complex ranking). This system was specifically designed to store tail pages and process tail queries. Various publications presented the architectures of open source search engines, such as Terrier [Ounis et al., 2006], Indri [Strohman et al., 2005a], and Lucene [McCandless et al., 2010]. Some open source search engines are listed in Table 4.2. A detailed comparison of open source search engines is available in Middleton and Baeza-Yates [2007].

The research on top-k query processing optimizations is vast. An early line of work tried to reduce the number of accumulators allocated and updated during the score computations. In Harman and Candela [1990], inverted lists are processed in decreasing order of their terms' IDF values. A page is fully scored only if it appears in the list of a high-IDF term. The remaining pages are not scored, reducing the overhead of sorting the accumulators. A similar idea was explored in Moffat and Zobel [1996], but this time explicitly bounding the number of allocated

Table 4.2: A list of open source query processing systems

Search engine	Description
ATIRE	Search engine (BSD license)
DataparkSearch	Website search engine (GNU GPL)
Elasticsearch	Search server based on Lucene (Apache license 2.0)
Galago	Search engine toolkit (BSD license)
Indri	Search engine (BSD-inspired license)
Lucene	Search engine library (Apache license 2.0)
MG4J	Distributed search engine (GNU LGPL)
mnoGoSearch	Website search engine (GNU GPL)
Solr	Distributed search engine based on Lucene (Apache license 2.0)
Seeks	User-centric search engine (Affero GPLv3)
Sphinx	Search engine (GNU GPLv2)
Terrier	Search engine (Mozilla public license)
Wumpus	Desktop search engine (GNU GPL)
Zettair	Search engine (BSD license)

accumulators. Another extension [Persin et al., 1996] was to decide on allocation and update of accumulators by comparisons between within-page term frequency values and some predetermined thresholds. In this case, the lists were frequency-sorted, allowing partial processing of the lists (rather than being entirely processed or omitted). A similar approach to that in Persin et al. [1996] was proposed in Anh et al. [2001], where the list entries were sorted in decreasing order of their contribution (impact) to page scores. The work in Anh and Moffat [2006] built on Anh et al. [2001] by proposing score-at-a-time processing of impact-sorted lists. Different from the earlier work in this line of research, this technique could guarantee a correct ranking compared to exhaustive evaluation. The technique was further improved in Strohman and Croft [2007], which combined accumulator pruning with inverted list skipping.

Another line of research proposed query processing optimizations for inverted lists whose entries are sorted in increasing document identifier order. The technique in Brown [1995] formed a candidate set including all pages in short inverted lists and previously stored high-scoring pages in long lists. The scores were computed only for these candidate pages, skipping all other pages in the inverted lists. Around the same time, the MaxScore optimization was proposed in Turtle and Flood [1995]. In this optimization, the lowest score in the min-heap and the maximum possible score contribution of each inverted list were used to compute certain score bounds. These bounds were used to identify pages that cannot make it into the heap and skip their score computation. The technique introduced in Strohman et al. [2005b] incorporated the candidate set idea proposed in Brown [1995] into the MaxScore optimization, enabling the computation of tighter score bounds early in the scoring process. Another important technique is the WAND optimization introduced in Broder et al. [2003a]. This optimization relies on a pointer movement strategy that allows many pages to be skipped without fully computing their scores. The WAND optimization was further improved in Ding and Suel [2011] using a block-max index, where the lists are divided into blocks and the maximum score contribution of each block was stored separately, enabling the skipping of entire blocks. In a concurrent work [Chakrabarti et al., 2011], a similar idea was applied to the MaxScore optimization. Improved versions of WAND and MaxScore were compared in Dimopoulos et al. [2013], again assuming a block-max index. More recent work proposed performing the pruning selectively, varying the aggressiveness of optimizations on a per query basis, based on efficiency and effectiveness predictors [Tonellotto et al., 2013] or current system load [Broccolo et al., 2013].

Yet another body of work proposed efficiency optimizations for the case where the score computations involve a query-independent component (e.g., a static page importance score) in addition to the usual query-dependent component. An early work here is Long and Suel [2003], which proposed various inverted list organizations (e.g., sorting inverted list entries in decreasing order of their page importance scores), accompanied with heuristics that prune list entires dynamically during the scoring computations. In Zhang et al. [2010], besides the page importance scores, additional query-independent information (e.g., the maximum possible score contribution the page can receive from any of its terms) was used when organizing the inverted list entries. This technique was shown to result in tighter score bounds than the technique in Long and Suel [2003], effectively facilitating early termination in score computations. In Shan et al. [2012], similar query processing optimizations were proposed for block-max indexes, assuming the presence of query-independent scores in the computation.

A number of studies evaluated the performance of distributed result retrieval systems under different assumptions [Cacheda et al., 2005, 2007, Cahoon et al., 2000]. A novel pipelined query processing technique was proposed for term-partitioned indexes in Moffat et al. [2007]. In this technique, a query is sequentially processed over the search nodes with intermediate rankings being communicated across the nodes according to a predetermined route. Although the pipelined scheme was shown to reduce the load imbalance in query processing at the batch level, the load imbalance problem remained at the micro scale due to random bursts in the query workload of search nodes. The pipelined query processing technique was improved later in Jonassen and Bratsberg [2012]. Several papers focused on reducing the hardware or energy costs of result retrieval systems. In Chowdhury and Pass [2003], a theoretical framework was proposed to minimize the hardware cost by optimizing the number of search clusters and the nodes per cluster, while trying to satisfy certain response time, throughput, and utilization constraints. In Freire et al. [2014], a self-adapting power model was proposed to adjust the number of active search nodes depending on the current query workload of the result retrieval system.

A large number of studies compared term-based and document-based index partitioning strategies, revealing the cons and pros of each strategy through simulations [Jeong and Omiecinski, 1995, Jonassen and Bratsberg, 2009, Ribeiro-Neto and Barbosa, 1998, Tomasic and Garcia-Molina, 1993b] or using actual retrieval systems [Badue et al., 2001, Cambazoglu et al., 2006, MacFarlane et al., 2000]. In the majority of these studies, document-based partitioning was found to lead to better query processing performance or scalability. A number of studies tried to improve the performance under term-based partitioning [Cambazoglu et al., 2013, Lucchese et al., 2007, Moffat et al., 2006, Zhang and Suel, 2007] while relatively few works focused on document-based partitioning [Badue et al., 2007, Ma et al., 2002, 2011]. Moreover, hybrid partitioning strategies that combine term- and document-based partitioning were proposed in Xi et al. [2002] and Feuerstein et al. [2009].

In the term-based index partitioning literature, load balancing was a common constraint while the performance objectives varied. In Moffat et al. [2006], a simple smallest-fit policy was adopted for load balancing a term-partitioned index. This heuristic is coupled with replication of frequently accessed inverted lists on the nodes. The strategy in Lucchese et al. [2007] assigned inverted lists that are co-accessed in queries to the same nodes, trying to optimize a performance objective that combines query processing throughput and average response time, each scaled by a relative importance factor. Also, a small fraction of frequently accessed inverted lists were replicated on all nodes. This strategy improved the locality of queries, i.e., on average, fewer nodes are involved in processing of a query. In Zhang and Suel [2007], inverted lists that have few pages in common were assigned to the same nodes. After obtaining an initial assignment of the lists via graph partitioning, different greedy heuristics were used to replicate selected inverted lists on a subset of search nodes. This approach aimed to reduce the communication overhead during the query processing. Along the same line, in Cambazoglu et al. [2013], the inverted index partitioning problem was modeled as a hypergraph partitioning problem. In this model, inverted lists that are concurrently accessed by many queries were assigned to the same search nodes. The model was shown to reduce the communication overhead while maintaining the computational load balance among the nodes, outperforming the strategies discussed above in terms of scalability.

Due to its simplicity, the work on document-based partitioning was more limited. In Ma et al. [2002], the authors proposed three different document-based partitioning strategies, two seeking storage balance and one seeking computational balance. The work demonstrated that document identifier reassignment and gap encoding can affect load balancing. In a latter work [Ma et al., 2011], the same authors proposed a document-based partitioning strategy that aimed to balance the storage overhead and computation load at the same time. Challenging the common belief that document-based partitioning leads to well-balanced query processing times, the experiments in Badue et al. [2007] indicated considerable load imbalance in processing of individual queries, mainly due to the variation in the disk caching behavior of the search nodes.

The roots of the selective search idea in web search engines go back to the collection selection problem in federated retrieval systems [Arguello et al., 2009, Callan et al., 1995, Lu and McKinley, 1999, Si and Callan, 2003]. An early work on collection selection is Callan et al. [1995], which proposed using inference networks to compute the relevance between queries and subcollections. The same problem was considered in Lu and McKinley [1999] in the context of partially replicated collections. In Si and Callan [2003], the relevance estimations were made by processing small indexes built on samples taken from each subcollection. A more recent work formulated the collection selection problem as a classification problem [Arguello et al., 2009].

The studies on collection selection assumed the availability of non-cooperative systems with readily partitioned collections. The studies on selective search architectures also tackled the problem of allocating the pages into subcollections [Ding et al., 2011, Kulkarni and Callan, 2010, Kulkarni et al., 2012, Puppin et al., 2010]. The work in Kulkarni and Callan [2010] evaluated random, source-based, and topic-based page allocation strategies. Among these, the topic-based allocation strategy was shown to couple better with the sampling-based collection selection strategy presented in Si and Callan [2003]. In a follow-up work [Kulkarni et al., 2012], the authors proposed three algorithms for ranking subcollections to be searched during the retrieval phase, assuming topical page collections and the same sampling-based selection strategy. In Puppin et al. [2010], a co-clustering algorithm was used to cluster the queries and pages simultaneously. In the retrieval phase, the similarity of a query to a collection is estimated jointly by first identifying the most similar query clusters for the query and then finding collections that are similar to these query clusters. The work in Ding et al. [2011] had a slightly different focus from the work discussed above in that it presented selective search strategies to increase the availability of pages and maintain the search result quality in case of failures or an overloaded search system.

Another line of research investigated static index pruning strategies [Altingovde et al., 2012, Blanco and Barreiro, 2007, 2010, Büttcher and Clarke, 2006, Carmel et al., 2001, Chen and Lee, 2013, de Moura et al., 2005, Ntoulas and Cho, 2007, Thota and Carterette, 2011]. In Carmel et al. [2001], the pruned index was obtained by removing from the original index the inverted list entries whose score contributions are lower than a certain threshold. Three different strategies were evaluated for setting the threshold: a uniform thresholding strategy that applies to all list entries and some term-specific thresholding strategies. The study in de Moura et al. [2005] argued that the technique in Carmel et al. [2001] is less effective in the case of conjunctive and phrase queries. A suitable technique was proposed by considering the co-occurrence of terms in the important sentences of pages. In this technique, the entry of a page is preserved in a pruned inverted list only if the respective term for the list appears in at least one of the important sentences of the page. Unlike the previous two approaches, the work in Büttcher and Clarke [2006] suggested keeping, in the pruned index, each page's most important terms, which are determined by a language model. In Blanco and Barreiro [2007], entire inverted lists were pruned based on their corresponding terms' informativeness. In Ntoulas and Cho [2007], different index pruning approaches were evaluated by removing the terms entirely, removing all inverted list entries belonging to a page, or removing selected inverted list entries. Some of the proposed pruning strategies could provide correctness guarantees for the results obtained by processing the pruned index with respect to the results that would be obtained by processing the full index. The work in Blanco and Barreiro [2010] applied the probability ranking principle to static index pruning.

A document-centric pruning approach that relies on the entropy measure was described in Thota and Carterette [2011]. The approach in Altingovde et al. [2012] made use of the term and page access statistics in query logs to guide the pruning process. In Chen and Lee [2013], it was argued that the simple pruning approach that uses global score thresholds [Carmel et al., 2001] is superior to all other pruning techniques in web-scale settings.

Despite its common use in practice, tiering took relatively less research attention [Baeza-Yates et al., 2009b, Leung et al., 2010, Risvik et al., 2003]. The tiering idea was first presented in Risvik et al. [2003], which assumed a three-tier system. The pages were allocated to tiers based on three different strategies, and the fall-through decisions were made by means of an ad-hoc formula. A different fall-through policy was proposed in Baeza-Yates et al. [2009b] for a two-tier search architecture. In this policy, for a selected subset of queries, the second tier is accessed in parallel with the first tier, rather than accessing the two tiers always sequentially. The computation in the second tier is halted early if the results retrieved from the first tier is found to be good quality. The work in Leung et al. [2010] investigated the page allocation problem in multi-tier search architectures. The tiering problem was formulated as an integer linear programming problem where the objective was to minimize the total penalty of retrieving the top documents of all queries from different tiers, each associated with a penalty.

The early work on caching in text retrieval systems focused on inverted list caching [Jónsson et al., 1998, Tomasic and Garcia-Molina, 1993a]. In Tomasic and Garcia-Molina [1993a], the authors use LRU as the eviction policy for a dynamic inverted list cache. In Jónsson et al. [1998], list replacement strategies were discussed in the context of a query refinement scenario assuming a single-user retrieval system. The first work to investigate query result caching in search engines was Markatos [2001], which provided a comparison between static and dynamic result caching approaches. Prefetching of query results was the focus of Lempel and Moran [2003]. The work aimed to predict, using offline statistics obtained from a query log, how many "next pages" should be cached for a query. Different strategies were presented in Ozcan et al. [2008] to fill a static result cache. The technique proposed in Baeza-Yates et al. [2007c] admitted queries to a dynamic result cache, selectively, based on a number of features. A hybrid result caching architecture, which involves a static part and a dynamic part, was proposed in Fagni et al. [2006]. The work also mentioned the possibility of maintaining a separate score cache. Two concurrent works investigated cost-aware result caching policies, where the execution cost of queries were taken into account when making the caching decisions [Altingovde et al., 2009, Gan and Suel, 2009]. The former work assumed a dynamic result cache, while the latter assumed a static result cache. The work in Sazoglu et al. [2013] went a step further and considered the estimated financial cost of queries when making the caching decisions.

In several studies, inverted list caching and result caching were considered together, occasionally with other types of caches [Baeza-Yates and Saint-Jean, 2003, Baeza-Yates et al., 2007b, Long and Suel, 2005, Saraiva et al., 2001]. A two-level cache architecture consisting of a dynamic inverted list cache and a dynamic result cache was used in Saraiva et al. [2001]. A similar study appears in Baeza-Yates and Saint-Jean [2003]; however, assuming static inverted list and result caches. The extensive study in Baeza-Yates et al. [2007b] covered many issues in static/dynamic result and list caching. The inverted list caching heuristic proposed by the authors took into account the size of inverted lists when making caching decisions (more specifically, the inverted lists are cached in decreasing order of their frequency/size ratios). In Long and Suel [2005], the two-level cache architecture in Saraiva et al. [2001] was extended to three levels by introducing a new cache that stores the intersections of inverted lists computed at query processing time. Two concurrent works proposed cache architectures with five levels for dynamic caching [Marin et al., 2010] and static caching [Ozcan et al., 2012], including additional caches such as a location cache and a page cache.

The work on caching is not limited to result and inverted list caching. For example, the experiments in Turpin et al. [2007] showed that retrieving pages from disk is the main cost in the snippet generation process and caching pages is essential for faster snippet generation. The dynamic page caching approach in Ceccarelli et al. [2011] stored only the sentences that are more likely to appear in snippets. In Ozcan et al. [2013], a result cache was coupled with a score cache in a dynamic caching setting. The impact of SSDs on the performance of inverted list, page, and snippet caches was investigated in Wang et al. [2013].

The freshness of search results received attention relatively recently. The first work on the topic is Cambazoglu et al. [2010a], which proposed a strategy to refresh cached query results using the idle cycles of the search backend. In a follow-up work [Jonassen et al., 2012], a predictive model was developed to decide which queries should be refreshed, aiming to optimize certain performance metrics such as response time and result freshness. The strategies in a concurrent line of research invalidated selected result cache entries based on the feedback obtained from the indexing system [Alici et al., 2011, Bai and Junqueira, 2012, Blanco et al., 2010, Bortnikov et al., 2011]. The work in Alici et al. [2012] proposed setting the time-to-live values of result cache entries, adaptively, on a per-query basis, instead of using the same time-to-live value for all entries.

The potential performance benefits of a multisite search architecture was first discussed in Cambazoglu et al. [2009]. A line of research investigated different query forwarding strategies [Baeza-Yates et al., 2009a, Cambazoglu et al., 2010b, Kayaaslan et al., 2011, Teymorian et al., 2013]. A novel query forwarding technique was introduced in Baeza-Yates et al. [2009a]. In this technique, each search site maintained the maximum possible score contribution of a term at each remote site. Given these precomputed scores, a local site could then compute an upper-bound on the scores each remote site can provide for a query. The forwarding decisions are made by comparisons between these upper bounds and the score of the bottom result in the local rankings. This technique was later improved by the linear programming solution introduced in Cambazoglu et al. [2010b], allowing tighter score bounds that increase the query locality. The impact of index updates on the score thresholds and the performance of the query forwarding algorithm introduced in Baeza-Yates et al. [2009a] was studied in Sarigiannis et al. [2009]. A multisite search engine setting with a fully replicated index on all sites was considered in Kayaaslan et al. [2011]. The queries were forwarded between search sites to reduce the electricity bill of a simulated search engine, exploiting the spatial and temporal variation in electricity prices. A similar approach was followed in Teymorian et al. [2013], but assuming a multisite search setting with a partitioned index.

In the context of multisite search architectures, a separate line of research investigated strategies for assigning web pages to search sites [Blanco et al., 2011, Brefeld et al., 2011, Junqueira et al., 2012, Kayaaslan et al., 2013]. A common objective in all of these strategies was to increase the locality of queries in a multisite search engine setting. In Brefeld et al. [2011], a machine learning model was built, using simple features such as the language, region, and size of the pages, to decide on which search sites each web page should be replicated. The work in Blanco et al. [2011] assigned each page uniquely to a single search site, called the master site for the page, using simple statistical evidence about the user interest in the page. Unlike the static page replication strategies, the strategy in Junqueira et al. [2012] made the replication decisions in an online manner and using local information. In Kayaaslan et al. [2013], three different replication algorithms were proposed to reduce the potential search quality loss, the average response time, or the query workload of a multisite search engine under certain capacity constraints. The potential gains that can be obtained by combining different page replication, query forwarding, and result caching strategies were investigated in Francès et al. [2014].

Various studies considered different architectural components or algorithmic optimizations in tandem, focusing on the interplay between them and the resulting trade-offs. In Puppin et al. [2010], an incremental caching strategy was proposed for collection selection architectures. The combination of result caching and static index pruning was investigated in Skobeltsyn et al. [2008], adopting the static index pruning strategies proposed in Ntoulas and Cho [2007]. In Lu and McKinley [2000], partial replica selection was coupled with result caching. The work in Tsegay et al. [2007] suggested caching inverted lists obtained after dynamic pruning, instead of full inverted lists. The effects of compression and document identifier reassignment on inverted list caching were investigated in Zhang et al. [2008] and Tonellotto et al. [2011], respectively.

A relatively recent line of research has focused on efficiency optimizations for machine-learned ranking. In Wang et al. [2010], a ranking model was learned in a way that will provide a good balance between search quality and efficiency. The follow-up work in Wang et al. [2011] proposed a cascaded ranking model, which involves increasingly complex ranking functions that are used to progressively prune the candidate set of pages. Several works tried to optimize the tree traversal in scoring computations. The work in Cambazoglu et al. [2010c] proposed four different early exit optimizations for second-phase ranking. These optimizations allowed short-circuiting score computations in ensembles of decision trees. Along the same line, cache-conscious optimizations were proposed in Tang et al. [2014] such that trees and candidate pages were processed in blocks. Architectural optimizations at the CPU level, aiming to reduce the number of control hazards and branch misprediction rate, were proposed in Asadi et al. [2014] and Lucchese et al. [2015]. As an application, machine learning was used in certain query processing optimizations. For example, it was applied to online query scheduling [Macdonald et al., 2012], predictive parallelization of queries [Jeon et al., 2014], and reducing extreme tail latency [Kim et al., 2015].

4.8   OPEN ISSUES IN QUERY PROCESSING

As we have seen in the previous section, there is a large body of research on architectural optimizations. However, this research has two shortcomings. First, most research is based on simulations or experiments on small-scale retrieval systems. Therefore, the research findings obtained so far may not always reflect the reality. The proposed optimizations need to be validated on real-life systems using large-scale data. Second, so far, most architectural optimizations are considered in isolation. An open research issue is to compare the performance of these architectures in a unifying study. Moreover, it may be possible to devise hybrid architectures that incorporate the existing architectural optimizations. For example, index partitioning, selective search, static index pruning, and tiering can all be coupled with each other in different ways to obtain further performance benefits.

So far, most efficiency research aimed to optimize key performance metrics such as response time and throughput. We believe that the research should also focus on a set of complementary metrics, such as the energy consumption, carbon footprint, and financial costs. More specifically, optimizing these metrics in the context of multisite search architectures is still an open research problem. Recent research already exploited certain properties of these architectures, such as the variation in energy prices, but others remain (e.g., exploiting the variation in weather conditions at different data center locations).

Finally, we believe that the efficiency optimizations, so far, had a system-centric perspective. The actual impact of these optimizations on the user experience is not well studied, with the exception of some recent works [Arapakis et al., 2014, Barreda-Ángeles et al., 2015]. This is mainly because system-centric optimizations and user-centric optimizations were often carried out separately. For example, an optimization can improve the existing response latency of a query processing system by 50 ms but, in most cases, it is hard to tell whether this would mean anything in terms of the end user experience. Therefore, an open problem is to unify the two concurrent lines of studies on system efficiency and user experience, by finding a conversion mechanism between the success measures used in these studies.

Table 4.3: A summary of the literature on query processing

Topic	Reference	Short description
	[Brin and Page, 1998]	Google (software focus)
	[Barroso et al., 2003]	Google (hardware focus)
Query	[Risvik et al., 2013]	Maguro
processing	[Ounis et al., 2006]	Terrier
system design	[Strohman et al., 2005a]	Indri
	[McCandless et al., 2010]	Lucene
	[Middleton and Baeza-Yates, 2007]	Comparison of open source search engines

Query processing optimizations	[Harman and Candela, 1990]	Processes only the lists of high-IDF terms
	[Moffat and Zobel, 1996]	Bounds accumulators (document-identifier-sorted lists)
	[Persin et al., 1996]	Bounds accumulators (frequency-sorted lists)
	[Anh et al., 2001]	Introduces impact-sorted lists
	[Anh and Moffat, 2006]	Score-at-a-time processing on impact-sorted lists
	[Strohman and Croft, 2007]	Combines accumulator pruning with skipping
	[Brown, 1995]	Scores only preselected candidate pages
	[Turtle and Flood, 1995]	MaxScore optimization
	[Strohman et al., 2005b]	Combines Brown [1995] and Turtle and Flood [1995]
	[Broder et al., 2003a]	WAND optimization
	[Ding and Suel, 2011]	Block-max index (applied to WAND)
	[Chakrabarti et al., 2011]	Similar to the block-max index (applied to MaxScore)
	[Dimopoulos et al., 2013]	Block-max index (applied to WAND and MaxScore)
	[Tonellotto et al., 2013]	Selective pruning using query performance predictors
	[Broccolo et al., 2013]	Selective pruning based on current system load
	[Long and Suel, 2003]	Optimizations in the case of query-independent scores
	[Zhang et al., 2010]	Optimizations improving Long and Suel [2003]
	[Shan et al., 2012]	Query-independent scores with block-max indexes
Parallel query processing	[Cahoon et al., 2000]	Simulations with varying query workloads
	[Cacheda et al., 2005]	Simulations with varying number of clusters and nodes
	[Cacheda et al., 2007]	Simulations with varying number of clusters and nodes
	[Moffat et al., 2007]	Pipelined query processing technique
	[Jonassen and Bratsberg, 2012]	Improves the pipelined query processing technique
	[Chowdhury and Pass, 2003]	Theoretical framework to minimize the hardware cost
	[Freire et al., 2014]	Energy saving based on query workload
Inverted Index Partitioning	[Tomasic and Garcia-Molina, 1993b]	Comparison on a simulated shared-nothing system
	[Jeong and Omiecinski, 1995]	Comparison on a simulated shared-everything system
	[Ribeiro-Neto and Barbosa, 1998]	Comparison on a simulated network of workstations
	[MacFarlane et al., 2000]	Comparison on distributed memory parallel computer
	[Badue et al., 2001]	Comparison on a network of workstations
	[Cambazoglu et al., 2006]	Comparison on a cluster of computers
	[Jonassen and Bratsberg, 2009]	Comparison on a simulated cluster of computers
	[Moffat et al., 2006]	Term-based (computational load balance)
	[Lucchese et al., 2007]	Term-based (throughput and average response time)
	[Zhang and Suel, 2007]	Term-based (communication volume)
	[Cambazoglu et al., 2013]	Term-based (communication volume)
	[Ma et al., 2002]	Document-based (storage/computational load balance)
	[Badue et al., 2007]	Document-based (computational load balance)
	[Ma et al., 2011]	Document-based (computational load balance)
	[Xi et al., 2002]	Hybrid strategy
	[Feuerstein et al., 2009]	Hybrid strategy

Selective search	[Callan et al., 1995]	Collection selection by inference networks (CORI)
	[Si and Callan, 2003]	Collection selection by page sampling (ReDDe))
	[Lu and McKinley, 1999]	Partial replica selection
	[Arguello et al., 2009]	Classification-based collection selection
	[Kulkarni and Callan, 2010]	Page allocation strategies
	[Kulkarni et al., 2012]	Collection ranking strategies
	[Puppin et al., 2010]	Co-clustering of queries and pages
	[Ding et al., 2011]	Increases the availability of pages
Static index pruning	[Carmel et al., 2001]	Term-centric pruning (relevance score thresholds)
	[de Moura et al., 2005]	Term-centric pruning (term co-occurrence in pages)
	[Büttcher and Clarke, 2006]	Document-centric pruning (KL divergence)
	[Blanco and Barreiro, 2007]	Prunes entire terms based on their informativeness
	[Ntoulas and Cho, 2007]	Pruning with correctness guarantees
	[Blanco and Barreiro, 2010]	Term-centric pruning (probability ranking principle)
	[Thota and Carterette, 2011]	Document-centric pruning (entropy)
	[Altingovde et al., 2012]	Term- and document-centric pruning (query views)
	[Chen and Lee, 2013]	Finds uniform pruning [Carmel et al., 2001] the best
Tiering	[Risvik et al., 2003]	Page allocation and fall-through policies
	[Baeza-Yates et al., 2009b]	Fall-through policy
	[Leung et al., 2010]	Page allocation policy
Caching	[Jónsson et al., 1998]	Dynamic inverted list caching
	[Tomasic and Garcia-Molina, 1993a]	Dynamic inverted list caching
	[Markatos, 2001]	Static and dynamic result caching
	[Lempel and Moran, 2003]	Dynamic result caching (prefetching next pages)
	[Ozcan et al., 2008]	Static result caching
	[Baeza-Yates et al., 2007c]	Dynamic result caching (admission)
	[Fagni et al., 2006]	Hybrid static and dynamic result caching
	[Gan and Suel, 2009]	Dynamic result caching (processing-cost-aware)
	[Altingovde et al., 2009]	Static result caching (processing-cost-aware)
	[Sazoglu et al., 2013]	Static and dynamic result caching (financial-cost-aware)
	[Saraiva et al., 2001]	Dynamic inverted list and result caching
	[Baeza-Yates and Saint-Jean, 2003]	Static inverted list and result caching
	[Long and Suel, 2005]	Dynamic inverted list, intersection, and result caching
	[Baeza-Yates et al., 2007b]	Static and dynamic inverted list and result caching
	[Marin et al., 2010]	Dynamic five-level cache
	[Ozcan et al., 2012]	Static five-level cache
	[Turpin et al., 2007]	Static and dynamic page caching
	[Ceccarelli et al., 2011]	Dynamic snippet caching
	[Ozcan et al., 2013]	Dynamic result and score cache
	[Wang et al., 2013]	Inverted list, page, and snippet cache (with SSDs)
Improving result freshness	[Cambazoglu et al., 2010a]	Refreshing (using simple features)
	[Jonassen et al., 2012]	Refreshing (using machine learning)
	[Blanco et al., 2010]	Invalidation
	[Bai and Junqueira, 2012]	Invalidation
	[Alici et al., 2011]	Invalidation (using timestamps of inverted lists)
	[Alici et al., 2012]	Adaptive time-to-live values

Multisite search architectures	[Cambazoglu et al., 2009]	Benefits of multisite search architectures
	[Baeza-Yates et al., 2009a]	Partitioned index, query forwarding (score thresholds)
	[Cambazoglu et al., 2010b]	Partitioned index, query forwarding (linear program)
	[Sarigiannis et al., 2009]	Effect of changing score thresholds on query forwarding
	[Kayaaslan et al., 2011]	Replicated index, query forwarding (energy prices)
	[Teymorian et al., 2013]	Partitioned index, query forwarding (energy prices)
	[Brefeld et al., 2011]	Page replication (uses a machine learning model)
	[Blanco et al., 2011]	Page assignment (selects a unique master site)
	[Junqueira et al., 2012]	Page replication (makes online and local decisions)
	[Kayaaslan et al., 2013]	Page replication (algorithmic perspective)
	[Francès et al., 2014]	Page replication, caching, and query forwarding
Trade-off analysis	[Puppin et al., 2010]	Incremental result caching and collection selection
	[Lu and McKinley, 2000]	Result caching and partial replica selection
	[Skobeltsyn et al., 2008]	Result caching and static index pruning
	[Tsegay et al., 2007]	Dynamic index pruning and inverted list caching
	[Zhang et al., 2008]	Compression and inverted list caching
	[Tonellotto et al., 2011]	Document identifier reassignment and list caching
Second-phase ranking efficiency	[Wang et al., 2010]	Learning model for effective and efficient ranking
	[Wang et al., 2011]	Cascaded learning model (progressively prunes pages)
	[Cambazoglu et al., 2010c]	Early exit optimizations for tree traversal
	[Tang et al., 2014]	Cache-conscious optimizations for tree traversal
	[Asadi et al., 2014]	CPU-level, architectural optimizations for tree traversal
	[Lucchese et al., 2015]	CPU-level, architectural tree traversal optimizations
Applications of machine learning	[Macdonald et al., 2012]	Latency prediction for online query scheduling
	[Jeon et al., 2014]	Models for selective parallelization of queries
	[Kim et al., 2015]	Models aiming to reduce extreme tail latency

CHAPTER 5

Concluding Remarks

In this book, we made an attempt to present the scalability and efficiency challenges in large-scale web search engines. In particular, we had a peek into the three main systems that are present in every web search engine architecture (web crawling, indexing, and query processing systems). When presenting each system, we tried to cover, as much as possible, the most prominent performance issues, employed efficiency solutions, related literature, and some open research problems.

We believe that, in the long term, the efficiency of web search engines will continue to attract the attention of researchers, both from the industry and academia. Unfortunately, we feel that a non-negligible fraction of the current academic research seem to deal with efficiency problems that are unrealistic or obsolete from the viewpoint of commercial search engines. We advocate that the research on efficiency should be driven by the actual needs of web search engines. Therefore, commercial search engine companies should reveal more of their efficiency problems and encourage research on those problems by providing external funding. As another handicap, we feel that most academic research is currently limited to simulations of web search systems and rely on very small or synthetic datasets. The search engine companies could help here as well, by making large-scale hardware infrastructures and real-life datasets more accessible to academic researchers.

We recommend keeping an eye on the advances in related research fields, such as databases, computer networks, distributed computing, and natural language processing. Existing solutions in these fields should be adapted to web search, instead of reinventing the wheel. The newly emerging techniques whose primary target is to improve the search quality should be closely followed and their implications for efficiency should be well understood. We also recommend observing the hardware trends. As the relative speeds of CPU, disk, RAM, and network continue to change over time, the nature of the performance bottlenecks and the feasibility of the existing solutions

Table 5.1: A summary of related books (WC: web crawling, I: indexing, QP: query processing)

Reference	Coverage of the book			Focus of the book	
	WC	I	QP	Area	Efficiency
[Witten et al., 1999]	None	High	Medium	Information retrieval	High
[Chakrabarti, 2002]	High	High	High	Web retrieval	High
[Grossman and Frieder, 2004]	None	Medium	High	Information retrieval	High
[Manning et al., 2008]	Low	High	High	Information retrieval	Low
[Croft et al., 2009]	Low	High	High	Web retrieval	Medium
[Chowdhury, 2010]	None	High	Medium	Library science	Low
[Büttcher et al., 2010]	Low	High	High	Information retrieval	High
[Baeza-Yates and Ribeiro-Neto, 2011]	High	High	High	Information retrieval	Medium

are likely to change as well.[1] Finally, the existing search algorithms need to be adapted to and evaluated on newly emerging computational architectures.

We find it somewhat difficult to speculate about the efficiency research in the distant future. But, we believe that certain limits will be reached at some point. For example, the increase in the query volumes will saturate as less and less people remain without Internet access. The growth rate of the Web will decrease as adoption of mobile devices increases and certain vertical apps become the primary means for information access. On the user side, ongoing performance improvements will decrease the tail search latency below the levels noticeable by the users. All of these may render further efficiency improvements less useful or critical. But, by the time this happens, the search business will have turned into a commodity that can be easily owned by many.

Before we finish our book, we would like to provide a brief survey of similar work. Although they are slightly outdated, the books with a greater focus on efficiency include Chakrabarti [2002], Grossman and Frieder [2004], and Witten et al. [1999]. Among the more recent books, Baeza-Yates and Ribeiro-Neto [2011] has the highest content coverage. Other recent books include Manning et al. [2008], Croft et al. [2009], and Chowdhury [2010]. Various properties of these books are summarized in Table 5.1. For shorter publications that discuss the efficiency challenges addressed in web search engines, the reader may refer to Henzinger et al. [2002], Baeza-Yates et al. [2007a], and Cambazoglu and Baeza-Yates [2011].

Finally, we would like the reader to know that we would be pleased to hear any kind of feedback about this book. Any comments regarding the book's comprehensiveness, readability issues, writing style, or references can be communicated to the authors via email at `barla@berk antbarlacambazoglu.com`. We thank the readers in advance.

---

[1]Two recent examples of this are the studies about the effect of flash SSDs on the performance of caching policies [Wang et al., 2013] and inverted index maintenance strategies [Jung et al., 2015].

Bibliography

E. Adar, J. Teevan, S. T. Dumais, and J. L. Elsas. The web changes everything: Understanding the dynamics of web content. In *Proceedings of the 2nd ACM International Conference on Web Search and Data Mining*, pages 282–291, New York, NY, USA, 2009. ISBN 978-1-60558-390-7. DOI: 10.1145/1498759.1498837. 25, 28

D. Ahlers and S. Boll. Adaptive geospatially focused crawling. In *Proceedings of the 18th ACM Conference on Information and Knowledge Management*, pages 445–454, New York, NY, USA, 2009. ISBN 978-1-60558-512-3. DOI: 10.1145/1645953.1646011. 26, 29

S. Alici, I. S. Altingovde, R. Ozcan, B. B. Cambazoglu, and O. Ulusoy. Timestamp-based result cache invalidation for web search engines. In *Proceedings of the 34th International ACM SIGIR Conference on Research and Development in Information Retrieval*, pages 973–982, New York, NY, USA, 2011. ISBN 978-1-4503-0757-4. DOI: 10.1145/2009916.2010046. 84, 88

S. Alici, I. S. Altingovde, R. Ozcan, B. B. Cambazoglu, and O. Ulusoy. Adaptive time-to-live strategies for query result caching in web search engines. In *Proceedings of the 34th European Conference on Advances in Information Retrieval*, pages 401–412, Berlin, Heidelberg, 2012. Springer-Verlag. ISBN 978-3-642-28996-5. DOI: 10.1007/978-3-642-28997-2_34. 84, 88

I. Altingovde, R. Ozcan, and O. Ulusoy. A cost-aware strategy for query result caching in web search engines. In M. Boughanem, C. Berrut, J. Mothe, and C. Soule-Dupuy, editors, *Advances in Information Retrieval*, volume 5478 of *Lecture Notes in Computer Science*, pages 628–636. Springer Berlin / Heidelberg, 2009. 83, 88

I. S. Altingovde, R. Ozcan, and O. Ulusoy. Static index pruning in web search engines: Combining term and document popularities with query views. *ACM Trans. Inf. Syst.*, 30(1):2:1–2:28, 2012. ISSN 1046-8188. DOI: 10.1145/2094072.2094074. 82, 83, 88

V. N. Anh and A. Moffat. Inverted index compression using word-aligned binary codes. *Inf. Retr.*, 8(1):151–166, 2005. ISSN 1386-4564. DOI: 10.1023/B:INRT.0000048490.99518.5c. 53, 56

V. N. Anh and A. Moffat. Pruned query evaluation using pre-computed impacts. In *Proceedings of the 29th Annual International ACM SIGIR Conference on Research and Development in Information Retrieval*, pages 372–379, New York, NY, USA, 2006. ISBN 1-59593-369-7. DOI: 10.1145/1148170.1148235. 79, 87

V. N. Anh, O. de Kretser, and A. Moffat. Vector-space ranking with effective early termination. In *Proceedings of the 24th Annual International ACM SIGIR Conference on Research and Development in Information Retrieval*, pages 35–42, New York, NY, USA, 2001. ISBN 1-58113-331-6. DOI: 10.1145/383952.383957. 79, 87

I. Arapakis, X. Bai, B. B. Cambazoglu, Impact of Response Latency on User Behavior in Web Search, *Proceedings of the 37th International ACM SIGIR Conference on Research and Development in Information Retrieval*, pages 103–112, New York, NY, USA, 2014. ISBN 978-1-4503-2257-7. DOI: 10.1145/2600428.2609627. 86

A. Arasu, J. Cho, H. Garcia-Molina, A. Paepcke, and S. Raghavan. Searching the Web. *ACM Trans. Internet Technol.*, 1(1):2–43, 2001. ISSN 1533-5399. DOI: 10.1145/383034.383035. 25, 27, 28

J. Arguello, J. Callan, and F. Diaz. Classification-based resource selection. In *Proceedings of the 18th ACM Conference on Information and Knowledge Management*, pages 1277–1286, New York, NY, USA, 2009. ISBN 978-1-60558-512-3. DOI: 10.1145/1645953.1646115. 81, 82, 88

D. Arroyuelo, S. González, M. Oyarzún, and V. Sepulveda. Document identifier reassignment and run-length-compressed inverted indexes for improved search performance. In *Proceedings of the 36th International ACM SIGIR Conference on Research and Development in Information Retrieval*, pages 173–182, New York, NY, USA, 2013. ISBN 978-1-4503-2034-4. DOI: 10.1145/2484028.2484079. 53, 57

N. Asadi, J. Lin, and A. P. de Vries. Runtime optimizations for tree-based machine learning models. *IEEE Trans. Knowledge Data Eng.*, 26(9):2281–2292, Sept 2014. ISSN 1041-4347. DOI: 10.1109/TKDE.2013.73. 85, 89

C. S. Badue, R. Baeza-Yates, B. Ribeiro-Neto, A. Ziviani, and N. Ziviani. Analyzing imbalance among homogeneous index servers in a web search system. *Inf. Process. Manage.*, 43(3):592–608, 2007. ISSN 0306-4573. DOI: 10.1016/j.ipm.2006.09.002. 81, 87

C. S. Badue, R. Baeza-Yates, B. A. Ribeiro-Neto, A. Ziviani, and N. Ziviani. Distributed query processing using partitioned inverted files. In *Proceedings of the 8th International Symposium on String Processing and Information Retrieval*, pages 10–20, 2001. ISBN 0-7695-1192-9. 81, 87

R. Baeza-Yates, C. Castillo, F. Junqueira, V. Plachouras, and F. Silvestri. Challenges on distributed web retrieval. In *Proceedings of the 23rd IEEE International Conference on Data Engineering*, pages 6–20, 2007a. DOI: 10.1109/ICDE.2007.367846. 92

R. Baeza-Yates and B. B. Cambazoglu. Scalability and efficiency challenges in large-scale web search engines. In *Proceedings of the Companion Publication of the 23rd International Conference on World Wide Web Companion*, pages 185–186, Republic and Canton of Geneva, Switzerland, 2014. International World Wide Web Conferences Steering Committee. ISBN 978-1-4503-2745-9. DOI: 10.1145/2567948.2577271. 2

R. Baeza-Yates and C. Castillo. Crawling the infinite web. *J. Web Eng.*, 6(1):49–72, March 2007. ISSN 1540-9589. 26, 29

R. Baeza-Yates and B. Ribeiro-Neto. *Modern Information Retrieval*. Addison-Wesley Publishing Company, USA, 2nd edition, 2011. ISBN 9780321416919. 92

R. Baeza-Yates and F. Saint-Jean. A three level search engine index based in query log distribution. In Mario A. Nascimento, Edleno S. de Moura, and Arlindo L. Oliveira, editors, *String Processing and Information Retrieval*, volume 2857 of *Lecture Notes in Computer Science*, pages 56–65. Springer Berlin / Heidelberg, 2003. DOI: 10.1007/978-3-540-39984-1_5. 83, 84, 88

R. Baeza-Yates, C. Castillo, M. Marin, and A. Rodriguez. Crawling a country: Better strategies than breadth-first for web page ordering. In *Special Interest Tracks and Posters of the 14th International Conference on World Wide Web*, pages 864–872, New York, NY, USA, 2005. ISBN 1-59593-051-5. DOI: 10.1145/1062745.1062768. 24, 28

R. Baeza-Yates, A. Gionis, F. Junqueira, V. Murdock, V. Plachouras, and F. Silvestri. The impact of caching on search engines. In *Proceedings of the 30th Annual International ACM SIGIR Conference on Research and Development in Information Retrieval*, pages 183–190, New York, NY, USA, 2007b. ISBN 978-1-59593-597-7. DOI: 10.1145/1277741.1277775. 83, 84, 88

R. Baeza-Yates, F. Junqueira, V. Plachouras, and H. Witschel. Admission policies for caches of search engine results. In Nivio Ziviani and Ricardo Baeza-Yates, editors, *String Processing and Information Retrieval*, volume 4726 of *Lecture Notes in Computer Science*, pages 74–85. Springer Berlin / Heidelberg, 2007c. DOI: 10.1007/978-3-540-75530-2_7. 83, 88

R. Baeza-Yates, A. Gionis, F. Junqueira, V. Plachouras, and L. Telloli. On the feasibility of multi-site web search engines. In *Proceedings of the 18th ACM Conference on Information and Knowledge Management*, pages 425–434, New York, NY, USA, 2009a. ISBN 978-1-60558-512-3. DOI: 10.1145/1645953.1646009. 84, 89

R. Baeza-Yates, V. Murdock, and C. Hauff. Efficiency trade-offs in two-tier web search systems. In *Proceedings of the 32nd International ACM SIGIR Conference on Research and Development in Information Retrieval*, pages 163–170, New York, NY, USA, 2009b. ISBN 978-1-60558-483-6. DOI: 10.1145/1571941.1571971. 83, 88

X. Bai and F. P. Junqueira. Online result cache invalidation for real-time web search. In *Proceedings of the 35th International ACM SIGIR Conference on Research and Development in Information Retrieval*, pages 641–650, New York, NY, USA, 2012. ISBN 978-1-4503-1472-5. DOI: 10.1145/2348283.2348369. 84, 88

X. Bai, B. B. Cambazoglu, and F. P. Junqueira. Discovering URLs through user feedback. In *Proceedings of the 20th ACM International Conference on Information and Knowledge Management*, pages 77–86, New York, NY, USA, 2011. ISBN 978-1-4503-0717-8. DOI: 10.1145/2063576.2063592. 24, 28

Z. Bar-Yossef, I. Keidar, and U. Schonfeld. Do not crawl in the DUST: Different URLs with similar text. In *Proceedings of the 16th International Conference on World Wide Web*, pages 111–120, New York, NY, USA, 2007. ISBN 978-1-59593-654-7. DOI: 10.1145/1242572.1242588. 26, 29

L. Barbosa and J. Freire. An adaptive crawler for locating hidden-web entry points. In *Proceedings of the 16th International Conference on World Wide Web*, pages 441–450, New York, NY, USA, 2007. ISBN 978-1-59593-654-7. DOI: 10.1145/1242572.1242632. 26, 29

L. Barbosa, A. C. Salgado, F. de Carvalho, J. Robin, and J. Freire. Looking at both the present and the past to efficiently update replicas of web content. In *Proceedings of the 7th Annual ACM International Workshop on Web Information and Data Management*, pages 75–80, New York, NY, USA, 2005. ISBN 1-59593-194-5. DOI: 10.1145/1097047.1097062. 25, 28

M. Barreda-Ángeles, I. Arapakis, X. Bai, B. B. Cambazoglu, and A. Pereda-Baños. Unconscious Physiological Effects of Search Latency on Users and Their Click Behaviour, *Proceedings of the 38th International ACM SIGIR Conference on Research and Development in Information Retrieval*, pages 203–212, New York, NY, USA, 2015. ISBN 978-1-4503-3621-7. DOI: 10.1145/2766462.2767719. 86

L. A. Barroso, J. Dean, and U. Hölzle. Web search for a planet: the Google cluster architecture. *IEEE Micro*, 23(2):22–28, 2003. ISSN 0272-1732. DOI: 10.1109/MM.2003.1196112. 78, 86

L. Becchetti, C. Castillo, D. Donato, R. Baeza-Yates, and S. Leonardi. Link analysis for web spam detection. *ACM Trans. Web*, 2(1):2:1–2:42, 2008. ISSN 1559-1131. DOI: 10.1145/1326561.1326563. 53, 56

K. Bharat, A. Broder, J. Dean, and M. R. Henzinger. A comparison of techniques to find mirrored hosts on the WWW. *J. Am. Soc. Inform. Sci.*, 51(12):1114–1122, 2000. ISSN 0002-8231. DOI: 10.1002/1097-4571(2000)9999:9999%3C::AID-ASI1025%3E3.0.CO;2-0. 26, 29

R. Blanco and A. Barreiro. TSP and cluster-based solutions to the reassignment of document identifiers. *Inform. Retr.*, 9(4):499–517, 2006. ISSN 1386-4564. DOI: 10.1007/s10791-006-6614-y. 53, 57

R. Blanco and A. Barreiro. Static pruning of terms in inverted files. In *Proceedings of the 29th European Conference on IR Research*, pages 64–75, Berlin, Heidelberg, 2007. Springer-Verlag. ISBN 978-3-540-71494-1. 82, 88

R. Blanco and A. Barreiro. Probabilistic static pruning of inverted files. *ACM Trans. Inf. Syst.*, 28(1):1:1–1:33, 2010. ISSN 1046-8188. DOI: 10.1145/1658377.1658378. 82, 88

R. Blanco, E. Bortnikov, F. Junqueira, R. Lempel, L. Telloli, and H. Zaragoza. Caching search engine results over incremental indices. In *Proceedings of the 33rd International ACM SIGIR Conference on Research and Development in Information Retrieval*, pages 82–89, New York, NY, USA, 2010. ISBN 978-1-4503-0153-4. DOI: 10.1145/1835449.1835466. 84, 88

R. Blanco, B. B. Cambazoglu, F. P. Junqueira, I. Kelly, and V. Leroy. Assigning documents to master sites in distributed search. In *Proceedings of the 20th ACM International Conference on Information and Knowledge Management*, pages 67–76, New York, NY, USA, 2011. ISBN 978-1-4503-0717-8. DOI: 10.1145/2063576.2063591. 85, 89

D. Blandford and G. Blelloch. Index compression through document reordering. In *Proceedings of the Data Compression Conference*, pages 342–351, Washington, DC, USA, 2002. IEEE Computer Society. 53, 57

P. Boldi, B. Codenotti, M. Santini, and S. Vigna. UbiCrawler: a scalable fully distributed web crawler. *Software Pract. Exper.*, 34(8):711–726, 2004. ISSN 1097-024X. DOI: 10.1002/spe.587. 24, 28

P. Boldi, A. Marino, M. Santini, and S. Vigna. BUbiNG: Massive crawling for the masses. In *Proceedings of the 23rd International Conference on World Wide Web*, pages 227–228, Republic and Canton of Geneva, Switzerland, 2014. International World Wide Web Conferences Steering Committee. ISBN 978-1-4503-2745-9. DOI: 10.1145/2567948.2577304. 24, 28

E. Bortnikov, R. Lempel, and K. Vornovitsky. Caching for realtime search. In *Proceedings of the 33rd European Conference on Advances in Information Retrieval*, pages 104–116, Berlin, Heidelberg, 2011. Springer-Verlag. ISBN 978-3-642-20160-8. URL http://dl.acm.org/citation.cfm?id=1996889.1996905. 84

U. Brefeld, B. B. Cambazoglu, and F. P. Junqueira. Document assignment in multi-site search engines. In *Proceedings of the 4th ACM International Conference on Web Search and Data Mining*, pages 575–584, New York, NY, USA, 2011. ISBN 978-1-4503-0493-1. DOI: 10.1145/1935826.1935907. 85, 89

S. Brin and L. Page. The anatomy of a large-scale hypertextual web search engine. *Comput. Netw. ISDN Syst.*, 30(1-7):107–117, 1998. ISSN 0169-7552. DOI: 10.1016/S0169-7552(98)00110-X. 23, 26, 28, 78, 86

D. Broccolo, C. Macdonald, S. Orlando, I. Ounis, R. Perego, F. Silvestri, and N. Tonellotto. Load-sensitive selective pruning for distributed search. In *Proceedings of the 22nd ACM International Conference on Conference on Information and Knowledge Management*, pages 379–388,

New York, NY, USA, 2013. ISBN 978-1-4503-2263-8. DOI: 10.1145/2505515.2505699. 80, 87

A. Z. Broder, S. C. Glassman, M. S. Manasse, and G. Zweig. Syntactic clustering of the Web. *Comput. Netw. ISDN Syst.*, 29(8-13):1157–1166, 1997. ISSN 0169-7552. DOI: 10.1016/S0169-7552(97)00031-7. 52, 56

A. Z. Broder, D. Carmel, M. Herscovici, A. Soffer, and J. Zien. Efficient query evaluation using a two-level retrieval process. In *Proceedings of the 12th International Conference on Information and Knowledge Management*, pages 426–434, New York, NY, USA, 2003a. ISBN 1-58113-723-0. DOI: 10.1145/956863.956944. 80, 87

A. Z. Broder, M. Najork, and J. L. Wiener. Efficient URL caching for World Wide Web crawling. In *Proceedings of the 12th International Conference on World Wide Web*, pages 679–689, New York, NY, USA, 2003b. ACM. ISBN 1-58113-680-3. DOI: 10.1145/775152.775247. 26, 29

E. W. Brown. Fast evaluation of structured queries for information retrieval. In *Proceedings of the 18th Annual International ACM SIGIR Conference on Research and Development in Information Retrieval*, pages 30–38, New York, NY, USA, 1995. ISBN 0-89791-714-6. DOI: 10.1145/215206.215329. 79, 87

E. W. Brown, J. P. Callan, and W. B. Croft. Fast incremental indexing for full-text information retrieval. In *Proceedings of the 20th International Conference on Very Large Data Bases*, pages 192–202, San Francisco, CA, USA, 1994. Morgan Kaufmann Publishers Inc. ISBN 1-55860-153-8. 54, 57

M. Burner. Crawling towards eternity: Building an archive of the World Wide Web. *Web Techniq. Mag.*, 2(5), 1997. 23, 28

S. Büttcher and C. L. Clarke. Hybrid index maintenance for contiguous inverted lists. *Inf. Retr.*, 11(3):175–207, 2008. ISSN 1386-4564. DOI: 10.1007/s10791-007-9042-8. 55, 57

S. Büttcher and C. L. A. Clarke. Indexing time vs. query time: Trade-offs in dynamic information retrieval systems. In *Proceedings of the 14th ACM International Conference on Information and Knowledge Management*, pages 317–318, New York, NY, USA, 2005. ISBN 1-59593-140-6. DOI: 10.1145/1099554.1099645. 54, 55, 57

S. Büttcher and C. L. A. Clarke. A document-centric approach to static index pruning in text retrieval systems. In *Proceedings of the 15th ACM International Conference on Information and Knowledge Management*, pages 182–189, New York, NY, USA, 2006. ISBN 1-59593-433-2. DOI: 10.1145/1183614.1183644. 82, 88

S. Büttcher, C. L. A. Clarke, and B. Lushman. Hybrid index maintenance for growing text collections. In *Proceedings of the 29th Annual International ACM SIGIR Conference on Research and Development in Information Retrieval*, pages 356–363, New York, NY, USA, 2006. ISBN 1-59593-369-7. DOI: 10.1145/1148170.1148233. 55, 57

S. Büttcher, C. L. A. Clarke, and G. V. Cormack. *Information Retrieval: Implementing and Evaluating Search Engines*. MIT Press, 2010. 92

F. Cacheda, V. Plachouras, and I. Ounis. A case study of distributed information retrieval architectures to index one terabyte of text. *Inf. Process. Manage.*, 41(5):1141–1161, 2005. ISSN 0306-4573. DOI: 10.1016/j.ipm.2004.05.002. 80, 87

F. Cacheda, V. Carneiro, V. Plachouras, and I. Ounis. Performance analysis of distributed information retrieval architectures using an improved network simulation model. *Inf. Process. Manage.*, 43(1):204–224, 2007. ISSN 0306-4573. DOI: 10.1016/j.ipm.2006.06.002. 80, 87

B. Cahoon, K. S. McKinley, and Z. Lu. Evaluating the performance of distributed architectures for information retrieval using a variety of workloads. *ACM Trans. Inf. Syst.*, 18(1):1–43, 2000. ISSN 1046-8188. DOI: 10.1145/333135.333136. 80, 87

R. Cai, J.-M. Yang, W. Lai, Y. Wang, and L. Zhang. iRobot: An intelligent crawler for web forums. In *Proceedings of the 17th International Conference on World Wide Web*, pages 447–456, New York, NY, USA, 2008. ISBN 978-1-60558-085-2. DOI: 10.1145/1367497.1367558. 26, 29

J. P. Callan, Z. Lu, and W. B. Croft. Searching distributed collections with inference networks. In *Proceedings of the 18th Annual International ACM SIGIR Conference on Research and Development in Information Retrieval*, pages 21–28, New York, NY, USA, 1995. ISBN 0-89791-714-6. DOI: 10.1145/215206.215328. 81, 88

B. B. Cambazoglu and R. Baeza-Yates. Scalability challenges in web search engines. In Massimo Melucci and Ricardo Baeza-Yates, editors, *Advanced Topics in Information Retrieval*, volume 33 of *The Information Retrieval Series*, pages 27–50. Springer Berlin Heidelberg, 2011. ISBN 978-3-642-20945-1. DOI: 10.1007/978-3-642-20946-8_2. 2, 92

B. B. Cambazoglu and R. Baeza-Yates. Scalability and efficiency challenges in commercial web search engines. In *Proceedings of the 36th International ACM SIGIR Conference on Research and Development in Information Retrieval*, pages 1124–1124, New York, NY, USA, 2013. ISBN 978-1-4503-2034-4. DOI: 10.1145/2484028.2484189. 2

B. B. Cambazoglu and R. Baeza-Yates. Scalability and efficiency challenges in large-scale web search engines. In *Proceedings of the 37th International ACM SIGIR Conference on Research and Development in Information Retrieval*, pages 1285–1285, New York, NY, USA, 2014. ISBN 978-1-4503-2257-7. DOI: 10.1145/2600428.2602291. 2

B. B. Cambazoglu and R. Baeza-Yates. Scalability and efficiency challenges in large-scale web search engines. In *Proceedings of the 8th ACM International Conference on Web Search and Data Mining*, pages 411–412, New York, NY, USA, 2015. ISBN 978-1-4503-3317-7. DOI: 10.1145/2684822.2697039. 2

B. B. Cambazoglu, A. Catal, and C. Aykanat. Effect of inverted index partitioning schemes on performance of query processing in parallel text retrieval systems. In Albert Levi, Erkay Savaş, Hüsnü Yenigün, Selim Balcısoy, and Yücel Saygın, editors, *Computer and Information Sciences – ISCIS 2006*, volume 4263 of *Lecture Notes in Computer Science*, pages 717–725. Springer Berlin Heidelberg, 2006. ISBN 978-3-540-47242-1. DOI: 10.1007/11902140_75. 81, 87

B. B. Cambazoglu, V. Plachouras, F. Junqueira, and L. Telloli. On the feasibility of geographically distributed web crawling. In *Proceedings of the 3rd International Conference on Scalable Information Systems*, pages 1–10, ICST, Brussels, Belgium, 2008. ICST (Institute for Computer Sciences, Social-Informatics and Telecommunications Engineering). ISBN 978-963-9799-28-8. 26, 28

B. B. Cambazoglu, V. Plachouras, and R. Baeza-Yates. Quantifying performance and quality gains in distributed web search engines. In *Proceedings of the 32nd International ACM SIGIR Conference on Research and Development in Information Retrieval*, pages 411–418, New York, NY, USA, 2009. ISBN 978-1-60558-483-6. DOI: 10.1145/1571941.1572013. 24, 28, 84, 89

B. B. Cambazoglu, F. P. Junqueira, V. Plachouras, S. Banachowski, B. Cui, S. Lim, and B. Bridge. A refreshing perspective of search engine caching. In *Proceedings of the 19th International Conference on World Wide Web*, pages 181–190, New York, NY, USA, 2010a. ACM. ISBN 978-1-60558-799-8. DOI: 10.1145/1772690.1772710. 84, 88

B. B. Cambazoglu, E. Varol, E. Kayaaslan, C. Aykanat, and R. Baeza-Yates. Query forwarding in geographically distributed search engines. In *Proceeding of the 33rd International ACM SIGIR Conference on Research and Development in Information Retrieval*, pages 90–97, New York, NY, USA, 2010b. ISBN 978-1-4503-0153-4. DOI: 10.1145/1835449.1835467. 84, 89

B. B. Cambazoglu, H. Zaragoza, O. Chapelle, J. Chen, C. Liao, Z. Zheng, and J. Degenhardt. Early exit optimizations for additive machine learned ranking systems. In *Proceedings of the 3rd ACM International Conference on Web Search and Data Mining*, pages 411–420, New York, NY, USA, 2010c. ISBN 978-1-60558-889-6. DOI: 10.1145/1718487.1718538. 85, 89

B. B. Cambazoglu, E. Kayaaslan, S. Jonassen, and C. Aykanat. A term-based inverted index partitioning model for efficient distributed query processing. *ACM Trans. Web*, 7(3):15:1–15:23, 2013. ISSN 1559-1131. DOI: 10.1145/2516633.2516637. 81, 87

D. Carmel, D. Cohen, R. Fagin, E. Farchi, M. Herscovici, Y. S. Maarek, and A. Soffer. Static index pruning for information retrieval systems. In *Proceedings of the 24th Annual International ACM SIGIR Conference on Research and Development in Information Retrieval*, pages 43–50, New York, NY, USA, 2001. ISBN 1-58113-331-6. DOI: 10.1145/383952.383958. 82, 83, 88

D. Ceccarelli, C. Lucchese, S. Orlando, R. Perego, and F. Silvestri. Caching query-biased snippets for efficient retrieval. In *Proceedings of the 14th International Conference on Extending Database Technology*, pages 93–104, New York, NY, USA, 2011. ISBN 978-1-4503-0528-0. DOI: 10.1145/1951365.1951379. 84, 88

K. Chakrabarti, S. Chaudhuri, and V. Ganti. Interval-based pruning for top-k processing over compressed lists. In *Proceedings of the 2011 IEEE 27th International Conference on Data Engineering*, pages 709–720, Washington, DC, USA, 2011. ISBN 978-1-4244-8959-6. DOI: 10.1109/ICDE.2011.5767855. 80, 87

S. Chakrabarti. *Mining the Web: Discovering Knowledge from Hypertext Data*. Morgan-Kauffman, 2002. ISBN ISBN 1-55860-754-4. 92

S. Chakrabarti, M. van den Berg, and B. Dom. Focused crawling: A new approach to topic-specific web resource discovery. *Comput. Netw.*, 31(11-16):1623–1640, 1999. ISSN 1389-1286. DOI: 10.1016/S1389-1286(99)00052-3. 26, 29

M. S. Charikar. Similarity estimation techniques from rounding algorithms. In *Proceedings of the 34th Annual ACM Symposium on Theory of Computing*, pages 380–388, New York, NY, USA, 2002. ISBN 1-58113-495-9. DOI: 10.1145/509907.509965. 52, 56

R.-C. Chen and C.-J. Lee. An information-theoretic account of static index pruning. In *Proceedings of the 36th International ACM SIGIR Conference on Research and Development in Information Retrieval*, pages 163–172, New York, NY, USA, 2013. ISBN 978-1-4503-2034-4. DOI: 10.1145/2484028.2484061. 82, 83, 88

Z. Cheng, B. Gao, C. Sun, Y. Jiang, and T.-Y. Liu. Let web spammers expose themselves. In *Proceedings of the 4th ACM International Conference on Web Search and Data Mining*, pages 525–534, New York, NY, USA, 2011. ISBN 978-1-4503-0493-1. DOI: 10.1145/1935826.1935902. 53, 56

F. Chierichetti, S. Lattanzi, F. Mari, and A. Panconesi. On placing skips optimally in expectation. In *Proceedings of the 1st International Conference on Web Search and Data Mining*, pages 15–24, New York, NY, USA, 2008. ACM. ISBN 978-1-59593-927-2. DOI: 10.1145/1341531.1341537. 52, 56

J. Cho and H. Garcia-Molina. Parallel crawlers. In *Proceedings of the 11th International Conference on World Wide Web*, pages 124–135, New York, NY, USA, 2002. ACM. ISBN 1-58113-449-5. DOI: 10.1145/511446.511464. 25, 28

J. Cho and H. Garcia-Molina. Effective page refresh policies for web crawlers. *ACM Trans. Database Syst.*, 28(4):390–426, 2003. ISSN 0362-5915. DOI: 10.1145/958942.958945. 25, 28

J. Cho and A. Ntoulas. Effective change detection using sampling. In *Proceedings of the 28th International Conference on Very Large Data Bases*, pages 514–525. VLDB Endowment, 2002. 25, 28

J. Cho, H. Garcia-Molina, and L. Page. Efficient crawling through URL ordering. *Computer Networks and ISDN Systems*, 30(1-7):161–172, 1998. ISSN 0169-7552. DOI: 10.1016/S0169-7552(98)00108-1. 24, 25, 28

J. Cho, N. Shivakumar, and H. Garcia-Molina. Finding replicated web collections. *ACM SIGMOD Record*, 29(2):355–366, 2000. ISSN 0163-5808. DOI: 10.1145/335191.335429. 26, 29

A. Chowdhury and G. Pass. Operational requirements for scalable search systems. In *Proceedings of the 12th International Conference on Information and Knowledge Management*, pages 435–442, New York, NY, USA, 2003. ACM. ISBN 1-58113-723-0. DOI: 10.1145/956863.956945. 80, 87

A. Chowdhury, O. Frieder, D. Grossman, and M. C. McCabe. Collection statistics for fast duplicate document detection. *ACM Trans. Inf. Syst.*, 20(2):171–191, 2002. ISSN 1046-8188. DOI: 10.1145/506309.506311. 52, 56

G. Chowdhury. *Introduction to Modern Information Retrieval*. Facet Publishing, 3rd edition, 2010. ISBN 185604694X, 9781856046947. 92

C. L. A. Clarke, G. V. Cormack, and F. J. Burkowski. Fast inverted indexes with on-line update. Technical report, CS-94-40, University of Waterloo, 1994. 54, 57

B. Croft, D. Metzler, and T. Strohman. *Search Engines: Information Retrieval in Practice*. Addison-Wesley Publishing Company, USA, 1st edition, 2009. ISBN 0136072240, 9780136072249. 92

D. Cutting and J. Pedersen. Optimization for dynamic inverted index maintenance. In *Proceedings of the 13th Annual International ACM SIGIR Conference on Research and Development in Information Retrieval*, pages 405–411, New York, NY, USA, 1990. ISBN 0-89791-408-2. DOI: 10.1145/96749.98245. 53, 55, 56, 57

A. Dasgupta, A. Ghosh, R. Kumar, C. Olston, S. Pandey, and A. Tomkins. The discoverability of the Web. In *Proceedings of the 16th International Conference on World Wide Web*, pages 421–430, New York, NY, USA, 2007. ACM. ISBN 978-1-59593-654-7. DOI: 10.1145/1242572.1242630. 24, 28

N. Daswani and M. Stoppelman. The anatomy of Clickbot.A. In *Proceedings of the 1st Workshop on Hot Topics in Understanding Botnets*, pages 11–11, Berkeley, CA, USA, 2007. USENIX Association. 53, 56

G. T. de Assis, A. H. Laender, M. A. Gonçalves, and A. S. da Silva. A genre-aware approach to focused crawling. *World Wide Web*, 12(3):285–319, 2009. ISSN 1386-145X. DOI: 10.1007/s11280-009-0063-7. 26, 29

E. S. de Moura, C. F. dos Santos, D. R. Fernandes, A. S. Silva, P. Calado, and M. A. Nascimento. Improving web search efficiency via a locality based static pruning method. In *Proceedings of the 14th International Conference on World Wide Web*, pages 235–244, New York, NY, USA, 2005. ACM. ISBN 1-59593-046-9. DOI: 10.1145/1060745.1060783. 82, 88

J. Dean. Challenges in building large-scale information retrieval systems: Invited talk. In *Proceedings of the 2nd ACM International Conference on Web Search and Data Mining*, pages 1–1, New York, NY, USA, 2009. ISBN 978-1-60558-390-7. DOI: 10.1145/1498759.1498761. 2

M. Diligenti, F. Coetzee, S. Lawrence, C. L. Giles, and M. Gori. Focused crawling using context graphs. In *Proceedings of the 26th International Conference on Very Large Data Bases*, pages 527–534, San Francisco, CA, USA, 2000. Morgan Kaufmann Publishers Inc. ISBN 1-55860-715-3. 26, 29

C. Dimopoulos, S. Nepomnyachiy, and T. Suel. Optimizing top-k document retrieval strategies for block-max indexes. In *Proceedings of the 6th ACM International Conference on Web Search and Data Mining*, pages 113–122, New York, NY, USA, 2013. ISBN 978-1-4503-1869-3. DOI: 10.1145/2433396.2433412. 80, 87

S. Ding and T. Suel. Faster top-k document retrieval using block-max indexes. In *Proceedings of the 34th International ACM SIGIR Conference on Research and Development in Information Retrieval*, pages 993–1002, New York, NY, USA, 2011. ISBN 978-1-4503-0757-4. DOI: 10.1145/2009916.2010048. 80, 87

S. Ding, J. Attenberg, and T. Suel. Scalable techniques for document identifier assignment in inverted indexes. In *Proceedings of the 19th International Conference on World Wide Web*, pages 311–320, New York, NY, USA, 2010. ACM. ISBN 978-1-60558-799-8. DOI: 10.1145/1772690.1772723. 53, 57

S. Ding, S. Gollapudi, S. Ieong, K. Kenthapadi, and A. Ntoulas. Indexing strategies for graceful degradation of search quality. In *Proceedings of the 34th International ACM SIGIR Conference on Research and Development in Information Retrieval*, pages 575–584, New York, NY, USA, 2011. ISBN 978-1-4503-0757-4. DOI: 10.1145/2009916.2009994. 82, 88

J. Edwards, K. McCurley, and J. Tomlin. An adaptive model for optimizing performance of an incremental web crawler. In *Proceedings of the 10th International Conference on World Wide Web*, pages 106–113, New York, NY, USA, 2001. ACM. ISBN 1-58113-348-0. DOI: 10.1145/371920.371960. 25, 28

D. Eichmann. The RBSE spider: balancing effective search against web load. In *Proceedings of the 1st World Wide Web Conference*, Geneva, Switzerland, 1994. 28

D. Eichmann. Ethical web agents. *Comput. Netw. ISDN Syst.*, 28(1-2):127–136, 1995. ISSN 0169-7552. DOI: 10.1016/0169-7552(95)00107-3. 26, 29

P. Elias. Universal codeword sets and representations of the integers. *IEEE Trans. Inf. Theor.*, 21 (2):194–203, 1975. ISSN 0018-9448. DOI: 10.1109/TIT.1975.1055349. 53, 56

P. Elias. Efficient storage and retrieval by content and address of static files. *J. ACM*, 21(2): 246–260, 1974. ISSN 0004-5411. DOI: 10.1145/321812.321820. 53, 56

J. Exposto, J. Macedo, A. Pina, A. Alves, and J. Rufino. Efficient partitioning strategies for distributed web crawling. In Teresa Vazão, Mário Freire, and Ilyoung Chong, editors, *Information Networking. Towards Ubiquitous Networking and Services*, volume 5200 of *Lecture Notes in Computer Science*, pages 544–553. Springer Berlin / Heidelberg, 2008. 26, 28

T. Fagni, R. Perego, F. Silvestri, and S. Orlando. Boosting the performance of web search engines: Caching and prefetching query results by exploiting historical usage data. *ACM Trans. Inform. Syst.*, 24(1):51–78, 2006. ISSN 1046-8188. DOI: 10.1145/1125857.1125859. 83, 88

C. Faloutsos. Signature files. In W. B. Frakes and R. Baeza-Yates, editors, *Information Retrieval: Data Structures and Algorithms*, pages 44–65. Prentice-Hall, Inc., Upper Saddle River, NJ, USA, 1992. ISBN 0-13-463837-9. 50, 56

R. M. Fano. On the number of bits required to implement an associative memory. *Memorandum*, 61, 1971. 53, 57

P. Ferragina and G. Manzini. On compressing the textual web. In *Proceedings of the 3rd ACM International Conference on Web Search and Data Mining*, pages 391–400, New York, NY, USA, 2010. ISBN 978-1-60558-889-6. DOI: 10.1145/1718487.1718536. 25, 28

D. Fetterly, M. Manasse, and M. Najork. Spam, damn spam, and statistics: Using statistical analysis to locate spam web pages. In *Proceedings of the 7th International Workshop on the Web and Databases: Colocated with ACM SIGMOD/PODS 2004*, pages 1–6, New York, NY, USA, 2004a. DOI: 10.1145/1017074.1017077. 52, 56

D. Fetterly, M. Manasse, M. Najork, and J. L. Wiener. A large-scale study of the evolution of web pages. *Software Pract. Exper.*, 34(2):213–237, 2004b. ISSN 0038-0644. DOI: 10.1002/spe.577. 25, 28

D. Fetterly, N. Craswell, and V. Vinay. The impact of crawl policy on web search effectiveness. In *Proceedings of the 32nd International ACM SIGIR Conference on Research and Development in Information Retrieval*, pages 580–587, New York, NY, USA, 2009. ISBN 978-1-60558-483-6. DOI: 10.1145/1571941.1572041. 25, 28

E. Feuerstein, M. Marin, M. Mizrahi, V. Gil-Costa, and R. Baeza-Yates. Two-dimensional distributed inverted files. In J. Karlgren, J. Tarhio, and H. Hyyrö, editors, *String Processing and Information Retrieval*, volume 5721 of *Lecture Notes in Computer Science*, pages 206–213. Springer Berlin Heidelberg, 2009. ISBN 978-3-642-03783-2. DOI: 10.1007/978-3-642-03784-9_20. 81, 87

M. Fontoura, E. Shekita, J. Y. Zien, S. Rajagopalan, and A. Neumann. High performance index build algorithms for intranet search engines. In *Proceedings of the 30th International Conference on Very Large Data Bases*, pages 1122–1133. VLDB Endowment, 2004. ISBN 0-12-088469-0. 50, 56

E. A. Fox and W. C. Lee. FAST-INV: A fast algorithm for building large inverted files. Technical report, TR-91-10, Virginia Polytechnic Institute and State University, 1991. 54, 57

G. Francès, X. Bai, B. B. Cambazoglu, and R. Baeza-Yates. Improving the efficiency of multi-site web search engines. In *Proceedings of the 7th ACM International Conference on Web Search and Data Mining*, pages 3–12, New York, NY, USA, 2014. ISBN 978-1-4503-2351-2. DOI: 10.1145/2556195.2556249. 85, 89

A. Freire, C. Macdonald, N. Tonellotto, I. Ounis, and F. Cacheda. A self-adapting latency/power tradeoff model for replicated search engines. In *Proceedings of the 7th ACM International Conference on Web Search and Data Mining*, pages 13–22, New York, NY, USA, 2014. ISBN 978-1-4503-2351-2. DOI: 10.1145/2556195.2556246. 80, 87

Q. Gan and T. Suel. Improved techniques for result caching in web search engines. In *Proceedings of the 18th International Conference on World Wide Web*, pages 431–440, New York, NY, USA, 2009. ISBN 978-1-60558-487-4. DOI: 10.1145/1526709.1526768. 83, 88

S. Golomb. Run-length encodings (corresp.). *IEEE Trans. Inf. Theor.*, 12(3):399–401, 1966. ISSN 0018-9448. DOI: 10.1109/TIT.1966.1053907. 53, 56

G. H. Gonnet, R. A. Baeza-Yates, and T. Snider. New indices for text: Pat trees and pat arrays. In W. B. Frakes and R. Baeza-Yates, editors, *Information Retrieval*, pages 66–82. Prentice-Hall, Inc., Upper Saddle River, NJ, USA, 1992. ISBN 0-13-463837-9. 50, 56

D. A. Grossman and O. Frieder. *Information Retrieval: Algorithms and Heuristics (The Kluwer International Series on Information Retrieval)*. Springer-Verlag New York, Inc., Secaucus, NJ, USA, 2004. ISBN 1402030037. 92

R. Guo, X. Cheng, H. Xu, and B. Wang. Efficient on-line index maintenance for dynamic text collections by using dynamic balancing tree. In *Proceedings of the 16th ACM Conference on Information and Knowledge Management*, pages 751–760, New York, NY, USA, 2007. ISBN 978-1-59593-803-9. DOI: 10.1145/1321440.1321545. 55, 57

S. Gurajada and S. K. Puligundla. On-line index maintenance using horizontal partitioning. In *Proceedings of the 18th ACM Conference on Information and Knowledge Management*, pages 435–444, New York, NY, USA, 2009. ISBN 978-1-60558-512-3. DOI: 10.1145/1645953.1646010. 54, 57

Z. Gyöngyi and H. Garcia-Molina. Link spam alliances. In *Proceedings of the 31st International Conference on Very Large Data Bases*, pages 517–528. VLDB Endowment, 2005. ISBN 1-59593-154-6. 26, 29

Z. Gyöngyi, H. Garcia-Molina, and J. Pedersen. Combating web spam with TrustRank. In *Proceedings of the 30th International Conference on Very Large Data Bases - Volume 30*, pages 576–587. VLDB Endowment, 2004. ISBN 0-12-088469-0. 52, 56

D. Harman and G. Candela. Retrieving records from a gigabyte of text on a mini-computer using statistical ranking. *J. Am. Soc. Inform. Sci.*, 41(8):581–589, 1990. 54, 57, 78, 87

D. Harman, R. Baeza-Yates, E. Fox, and W. Lee. Inverted files. In W. B. Frakes and R. Baeza-Yates, editors, *Information Retrieval: Data Structures and Algorithms*, pages 28–43. Prentice-Hall, Inc., Upper Saddle River, NJ, USA, 1992. ISBN 0-13-463837-9. 50, 56

S. Heinz and J. Zobel. Efficient single-pass index construction for text databases. *J. Am. Soc. Inform. Sci.*, 54(8):713–729, 2003. ISSN 1532-2882. DOI: 10.1002/asi.10268. 54, 57

M. Henzinger. Finding near-duplicate web pages: A large-scale evaluation of algorithms. In *Proceedings of the 29th Annual International ACM SIGIR Conference on Research and Development in Information Retrieval*, pages 284–291, New York, NY, USA, 2006. ISBN 1-59593-369-7. DOI: 10.1145/1148170.1148222. 52, 56

M. R. Henzinger, R. Motwani, and C. Silverstein. Challenges in web search engines. *SIGIR Forum*, 36(2):11–22, 2002. ISSN 0163-5840. DOI: 10.1145/792550.792553. 92

A. Heydon and M. Najork. Mercator: A scalable, extensible web crawler. *World Wide Web*, 2(4): 219–229, 1999. ISSN 1386-145X. DOI: 10.1023/A:1019213109274. 23, 28

P. Heymann, G. Koutrika, and H. Garcia-Molina.   Can social bookmarking improve web search?   In *Proceedings of the 1st International Conference on Web Search and Data Mining*, pages 195–206, New York, NY, USA, 2008. ACM.   ISBN 978-1-59593-927-2.   DOI: 10.1145/1341531.1341558. 24, 28

J. Hirai, S. Raghavan, H. Garcia-Molina, and A. Paepcke. WebBase: a repository of web pages. In *Proceedings of the 9th International Conference on World Wide Web*, pages 277–293, Amsterdam, The Netherlands, The Netherlands, 2000. North-Holland Publishing Co.   DOI: 10.1016/S1389-1286(00)00063-3. 25, 28

B.-J. (Paul) Hsu and G. Ottaviano. Space-efficient data structures for top-k completion. In *Proceedings of the 22nd International Conference on World Wide Web*, pages 583–594, Republic and Canton of Geneva, Switzerland, 2013. International World Wide Web Conferences Steering Committee. ISBN 978-1-4503-2035-1. 52, 56

M. Jeon, S. Kim, S.-w. Hwang, Y. He, S. Elnikety, A. L. Cox, and S. Rixner. Predictive parallelization: Taming tail latencies in web search. In *Proceedings of the 37th International ACM SIGIR Conference on Research and Development in Information Retrieval*, pages 253–262, New York, NY, USA, 2014. ISBN 978-1-4503-2257-7. DOI: 10.1145/2600428.2609572. 86, 89

B.-S. Jeong and E. Omiecinski. Inverted file partitioning schemes in multiple disk systems. *IEEE Trans. Parallel Distrib. Syst.*, 6(2):142–153, 1995. ISSN 1045-9219. DOI: 10.1109/71.342125. 80, 87

S. Jonassen and S. E. Bratsberg. Impact of the query model and system settings on performance of distributed inverted indexes. In *Norsk Informatikkonferance 2009*, pages 143–154, 2009. 81, 87

S. Jonassen and S. E. Bratsberg. Improving the performance of pipelined query processing with skipping. In *Proceedings of the 13th International Conference on Web Information Systems Engineering*, pages 1–15, Berlin, Heidelberg, 2012. Springer-Verlag. ISBN 978-3-642-35062-7. DOI: 10.1007/978-3-642-35063-4_1. 80, 87

S. Jonassen, B. B. Cambazoglu, and F. Silvestri. Prefetching query results and its impact on search engines. In *Proceedings of the 35th International ACM SIGIR Conference on Research and Development in Information Retrieval*, pages 631–640, New York, NY, USA, 2012. ISBN 978-1-4503-1472-5. DOI: 10.1145/2348283.2348368. 84, 88

B. T. Jónsson, M. J. Franklin, and D. Srivastava. Interaction of query evaluation and buffer management for information retrieval. *ACM SIGMOD Record*, 27(2):118–129, 1998. ISSN 0163-5808. DOI: 10.1145/276305.276316. 83, 88

W. Jung, H. Roh, M. Shin, and S. Park. Inverted index maintenance strategy for flashSSDs. *Inf. Syst.*, 49(C):25–39, 2015. ISSN 0306-4379. DOI: 10.1016/j.is.2014.11.004. 54, 57, 92

F. P. Junqueira, V. Leroy, and M. Morel. Reactive index replication for distributed search engines. In *Proceedings of the 35th International ACM SIGIR Conference on Research and Development in Information Retrieval*, pages 831–840, New York, NY, USA, 2012. ISBN 978-1-4503-1472-5. DOI: 10.1145/2348283.2348394. 85, 89

E. Kayaaslan, B. B. Cambazoglu, R. Blanco, F. P. Junqueira, and C. Aykanat. Energy-price-driven query processing in multi-center web search engines. In *Proceedings of the 34th International ACM SIGIR Conference on Research and Development in Information Retrieval*, pages 983–992, New York, NY, USA, 2011. ISBN 978-1-4503-0757-4. DOI: 10.1145/2009916.2010047. 84, 85, 89

E. Kayaaslan, B. B. Cambazoglu, and C. Aykanat. Document replication strategies for geographically distributed web search engines. *Inf. Process. Manage.*, 49(1):51–66, 2013. ISSN 0306-4573. DOI: 10.1016/j.ipm.2012.01.002. 85, 89

S. Kim, Y. He, S.-w. Hwang, S. Elnikety, and S. Choi. Delayed-dynamic-selective (DDS) prediction for reducing extreme tail latency in web search. In *Proceedings of the 8th ACM International Conference on Web Search and Data Mining*, pages 7–16, New York, NY, USA, 2015. ISBN 978-1-4503-3317-7. DOI: 10.1145/2684822.2685289. 86, 89

R. Konow, G. Navarro, C. L. A. Clarke, and A. López-Ortíz. Faster and smaller inverted indices with treaps. In *Proceedings of the 36th International ACM SIGIR Conference on Research and Development in Information Retrieval*, pages 193–202, New York, NY, USA, 2013. ISBN 978-1-4503-2034-4. DOI: 10.1145/2484028.2484088. 52, 56

H. S. Koppula, K. P. Leela, A. Agarwal, K. P. Chitrapura, S. Garg, and A. Sasturkar. Learning URL patterns for webpage de-duplication. In *Proceedings of the 3rd ACM International Conference on Web Search and Data Mining*, pages 381–390, New York, NY, USA, 2010. ISBN 978-1-60558-889-6. DOI: 10.1145/1718487.1718535. 26, 29

V. Krishnan and R. Raj. Web spam detection with anti-trust rank. In *Proceedings of the 2nd International Workshop on Adversarial Information Retrieval on the Web*, pages 37–40, 2006. 52, 56

A. Kulkarni and J. Callan. Document allocation policies for selective searching of distributed indexes. In *Proceedings of the 19th ACM International Conference on Information and Knowledge Management*, pages 449–458, New York, NY, USA, 2010. ISBN 978-1-4503-0099-5. DOI: 10.1145/1871437.1871497. 82, 88

A. Kulkarni, A. S. Tigelaar, D. Hiemstra, and J. Callan. Shard ranking and cutoff estimation for topically partitioned collections. In *Proceedings of the 21st ACM International Conference on Information and Knowledge Management*, pages 555–564, New York, NY, USA, 2012. ISBN 978-1-4503-1156-4. DOI: 10.1145/2396761.2396833. 82, 88

H.-T. Lee, D. Leonard, X. Wang, and D. Loguinov.   IRLbot: scaling to 6 billion pages and beyond.   In *Proceedings of the 17th International Conference on World Wide Web*, pages 427–436, New York, NY, USA, 2008. ACM.   ISBN 978-1-60558-085-2.   DOI: 10.1145/1367497.1367556. 24, 28

T. Lee, J. Kim, J. W. Kim, S.-R. Kim, and K. Park.   Detecting soft errors by redirection classification.   In *Proceedings of the 18th International Conference on World Wide Web*, pages 1119–1120, New York, NY, USA, 2009. ACM.   ISBN 978-1-60558-487-4.   DOI: 10.1145/1526709.1526886. 26, 29

D. Lefortier, L. Ostroumova, E. Samosvat, and P. Serdyukov.   Timely crawling of high-quality ephemeral new content. In *Proceedings of the 22nd ACM International Conference on Information and Knowledge Management*, pages 745–750, New York, NY, USA, 2013. ISBN 978-1-4503-2263-8. DOI: 10.1145/2505515.2505641. 24, 28

R. Lempel and S. Moran.   Predictive caching and prefetching of query results in search engines. In *Proceedings of the 12th International Conference on World Wide Web*, pages 19–28, New York, NY, USA, 2003. ACM. ISBN 1-58113-680-3. DOI: 10.1145/775152.775156. 83, 88

N. Lester, J. Zobel, and H. E. Williams. In-place versus re-build versus re-merge: Index maintenance strategies for text retrieval systems. In *Proceedings of the 27th Australasian Conference on Computer Science - Volume 26*, pages 15–23, Darlinghurst, New South Wales, Australia, 2004. Australian Computer Society, Inc. 54, 57

N. Lester, J. Zobel, and H. Williams.   Efficient online index maintenance for contiguous inverted lists.   *Inf. Process. Manage.*, 42(4):916–933, July 2006.   ISSN 0306-4573.   DOI: 10.1016/j.ipm.2005.09.005. 54, 57

N. Lester, A. Moffat, and J. Zobel.   Efficient online index construction for text databases. *ACM Trans. Database Syst.*, 33(3):1–33, 2008. ISSN 0362-5915. DOI: 10.1145/1386118.1386125. 54, 57

G. Leung, N. Quadrianto, K. Tsioutsiouliklis, and A. J. Smola.   Optimal web-scale tiering as a flow problem. In J. D. Lafferty, C. K. I. Williams, J. Shawe-Taylor, R. S. Zemel, and A. Culotta, editors, *Advances in Neural Information Processing Systems 23*, pages 1333–1341. Curran Associates, Inc., 2010. 83, 88

J.-L. Lin. Detection of cloaked web spam by using tag-based methods. *Expert Syst. Appl.*, 36(4): 7493–7499, 2009. ISSN 0957-4174. DOI: 10.1016/j.eswa.2008.09.056. 53, 56

M. Liu, R. Cai, M. Zhang, and L. Zhang.   User browsing behavior-driven web crawling.   In *Proceedings of the 20th ACM International Conference on Information and Knowledge Management*, pages 87–92, New York, NY, USA, 2011.   ISBN 978-1-4503-0717-8.   DOI: 10.1145/2063576.2063593. 25, 28

Y. Liu, F. Chen, W. Kong, H. Yu, M. Zhang, S. Ma, and L. Ru. Identifying web spam with the wisdom of the crowds. *ACM Trans. Web*, 6(1):2:1–2:30, 2012. ISSN 1559-1131. DOI: 10.1145/2109205.2109207. 53, 56

X. Long and T. Suel. Optimized query execution in large search engines with global page ordering. In *Proceedings of the 29th International Conference on Very Large Data Bases*, pages 129–140. VLDB Endowment, 2003. ISBN 0-12-722442-4. 80, 87

X. Long and T. Suel. Three-level caching for efficient query processing in large web search engines. In *Proceedings of the 14th International Conference on World Wide Web*, pages 257–266, New York, NY, USA, 2005. ACM. ISBN 1-59593-046-9. DOI: 10.1145/1060745.1060785. 83, 84, 88

Z. Lu and K. S. McKinley. Partial replica selection based on relevance for information retrieval. In *Proceedings of the 22nd Annual International ACM SIGIR Conference on Research and Development in Information Retrieval*, pages 97–104, New York, NY, USA, 1999. ISBN 1-58113-096-1. DOI: 10.1145/312624.312662. 81, 82, 88

Z. Lu and K. S. McKinley. Partial collection replication versus caching for information retrieval systems. In *Proceedings of the 23rd Annual International ACM SIGIR Conference on Research and Development in Information Retrieval*, pages 248–255, New York, NY, USA, 2000. ISBN 1-58113-226-3. DOI: 10.1145/345508.345591. 85, 89

C. Lucchese, S. Orlando, R. Perego, and F. Silvestri. Mining query logs to optimize index partitioning in parallel web search engines. In *Proceedings of the 2nd International Conference on Scalable Information Systems*, pages 1–9, ICST, Brussels, Belgium, Belgium, 2007. ICST (Institute for Computer Sciences, Social-Informatics and Telecommunications Engineering). ISBN 978-1-59593-757-5. 81, 87

C. Lucchese, F. M. Nardini, S. Orlando, R. Perego, N. Tonellotto, and R. Venturini. Quickscorer: A fast algorithm to rank documents with additive ensembles of regression trees. In *Proceedings of the 38th International ACM SIGIR Conference on Research and Development in Information Retrieval*, pages 73–82, New York, NY, USA, 2015. ISBN 978-1-4503-3621-5. DOI: 10.1145/2766462.2767733. 85, 89

Y.-C. Ma, T.-F. Chen, and C.-P. Chung. Posting file partitioning and parallel information retrieval. *J. Syst. Softw.*, 63(2):113–127, 2002. ISSN 0164-1212. DOI: 10.1016/S0164-1212(01)00119-4. 81, 87

Y.-C. Ma, C.-P. Chung, and T.-F. Chen. Load and storage balanced posting file partitioning for parallel information retrieval. *J. Syst. Softw.*, 84(5):864–884, 2011. ISSN 0164-1212. DOI: 10.1016/j.jss.2011.01.028. 81, 87

C. Macdonald, N. Tonellotto, and I. Ounis. Learning to predict response times for online query scheduling. In *Proceedings of the 35th International ACM SIGIR Conference on Research and Development in Information Retrieval*, pages 621–630, New York, NY, USA, 2012. ISBN 978-1-4503-1472-5. DOI: 10.1145/2348283.2348367. 86, 89

A. MacFarlane, J. A. McCann, and S. E. Robertson. Parallel search using partitioned inverted files. In *Proceedings of the 7th International Symposium on String Processing and Information Retrieval*, pages 209–220, 2000. DOI: 10.1109/SPIRE.2000.878197. 81, 87

U. Manber and G. Myers. Suffix arrays: A new method for on-line string searches. In *Proceedings of the 1st Annual ACM-SIAM Symposium on Discrete Algorithms*, pages 319–327, Philadelphia, PA, USA, 1990. Society for Industrial and Applied Mathematics. ISBN 0-89871-251-3. 50

G. S. Manku, A. Jain, and A. D. Sarma. Detecting near-duplicates for web crawling. In *Proceedings of the 16th International Conference on World Wide Web*, pages 141–150, New York, NY, USA, 2007. ACM. ISBN 978-1-59593-654-7. DOI: 10.1145/1242572.1242592. 52, 56

C. D. Manning, P. Raghavan, and H. Schütze. *Introduction to Information Retrieval*. Cambridge University Press, New York, NY, USA, 2008. ISBN 0521865719, 9780521865715. 92

G. Margaritis and S. V. Anastasiadis. Low-cost management of inverted files for online full-text search. In *Proceedings of the 18th ACM Conference on Information and Knowledge Management*, pages 455–464, New York, NY, USA, 2009. ISBN 978-1-60558-512-3. DOI: 10.1145/1645953.1646012. 55, 57

M. Marin, V. Gil-Costa, and C. Gomez-Pantoja. New caching techniques for web search engines. In *Proceedings of the 19th ACM International Symposium on High Performance Distributed Computing*, pages 215–226, New York, NY, USA, 2010. ISBN 978-1-60558-942-8. DOI: 10.1145/1851476.1851502. 84, 88

E. P. Markatos. On caching search engine query results. *Comput. Commun.*, 24(2):137–143, 2001. ISSN 0140-3664. DOI: DOI: 10.1016/S0140-3664(00)00308-X. 83, 88

M. McCandless, E. Hatcher, and O. Gospodnetic. *Lucene in Action, Second Edition: Covers Apache Lucene 3.0*. Manning Publications Co., Greenwich, CT, USA, 2010. ISBN 1933988177, 9781933988177. 78, 86

R. Mccreadie, C. Macdonald, and I. Ounis. MapReduce indexing strategies: Studying scalability and efficiency. *Inf. Process. Manage.*, 48(5):873–888, 2012. ISSN 0306-4573. DOI: 10.1016/j.ipm.2010.12.003. 54, 57

S. Melnik, S. Raghavan, B. Yang, and H. Garcia-Molina. Building a distributed full-text index for the Web. *ACM Trans. Inform. Syst.*, 19(3):217–241, 2001. ISSN 1046-8188. DOI: 10.1145/502115.502116. 54, 57

C. Middleton and R. Baeza-Yates. A comparison of open source search engines, 2007. 78, 86

A. Moffat and T. A. H. Bell. In situ generation of compressed inverted files. *J. Am. Soc. Inform. Sci.*, 46(7):537–550, 1995. ISSN 0002-8231. DOI: 10.1002/(SICI)1097-4571(199508)46:7%3C537::AID-ASI7%3E3.0.CO;2-P. 54, 57

A. Moffat and L. Stuiver. Binary interpolative coding for effective index compression. *Inform. Retri.*, 3(1):25–47, 2000. ISSN 1386-4564. DOI: 10.1023/A:1013002601898. 53, 56

A. Moffat and J. Zobel. Self-indexing inverted files for fast text retrieval. *ACM Trans. Inform. Syst.*, 14(4):349–379, 1996. ISSN 1046-8188. DOI: 10.1145/237496.237497. 52, 56, 78, 87

A. Moffat, W. Webber, and J. Zobel. Load balancing for term-distributed parallel retrieval. In *Proceedings of the 29th Annual International ACM SIGIR Conference on Research and Development in Information Retrieval*, pages 348–355, New York, NY, USA, 2006. ISBN 1-59593-369-7. DOI: 10.1145/1148170.1148232. 81, 87

A. Moffat, W. Webber, J. Zobel, and R. Baeza-Yates. A pipelined architecture for distributed text query evaluation. *Inf. Retr.*, 10(3):205–231, 2007. ISSN 1386-4564. DOI: 10.1007/s10791-006-9014-4. 80, 87

M. Najork. Detecting quilted web pages at scale. In *Proceedings of the 35th International ACM SIGIR Conference on Research and Development in Information Retrieval*, pages 385–394, New York, NY, USA, 2012. ISBN 978-1-4503-1472-5. DOI: 10.1145/2348283.2348337. 52, 56

M. Najork and J. L. Wiener. Breadth-first crawling yields high-quality pages. In *Proceedings of the 10th International Conference on World Wide Web*, pages 114–118, New York, NY, USA, 2001. ACM. ISBN 1-58113-348-0. DOI: 10.1145/371920.371965. 24, 28

A. Ntoulas and J. Cho. Pruning policies for two-tiered inverted index with correctness guarantee. In *Proceedings of the 30th Annual International ACM SIGIR Conference on Research and Development in Information Retrieval*, pages 191–198, New York, NY, USA, 2007. ISBN 978-1-59593-597-7. DOI: 10.1145/1277741.1277776. 82, 85, 88

A. Ntoulas, J. Cho, and C. Olston. What's new on the Web?: the evolution of the Web from a search engine perspective. In *Proceedings of the 13th International Conference on World Wide Web*, pages 1–12, New York, NY, USA, 2004. ACM. ISBN 1-58113-844-X. DOI: 10.1145/988672.988674. 25, 28

A. Ntoulas, P. Zerfos, and J. Cho. Downloading textual hidden web content through keyword queries. In *Proceedings of the 5th ACM/IEEE-CS Joint Conference on Digital Libraries*, pages 100–109, New York, NY, USA, 2005. ISBN 1-58113-876-8. DOI: 10.1145/1065385.1065407. 26, 29

A. Ntoulas, M. Najork, M. Manasse, and D. Fetterly. Detecting spam web pages through content analysis. In *Proceedings of the 15th International Conference on World Wide Web*, pages 83–92, New York, NY, USA, 2006. ACM. ISBN 1-59593-323-9. DOI: 10.1145/1135777.1135794. 52, 56

C. Olston and M. Najork. Web crawling. *Found. Trends Inf. Retr.*, 4(3):175–246, 2010. ISSN 1554-0669. DOI: 10.1561/1500000017. 26, 27

C. Olston and S. Pandey. Recrawl scheduling based on information longevity. In *Proceedings of the 17th International Conference on World Wide Web*, pages 437–446, New York, NY, USA, 2008. ACM. ISBN 978-1-60558-085-2. DOI: 10.1145/1367497.1367557. 25, 28

G. Ottaviano and R. Venturini. Partitioned Elias-Fano indexes. In *Proceedings of the 37th International ACM SIGIR Conference on Research and Development in Information Retrieval*, pages 273–282, New York, NY, USA, 2014. ISBN 978-1-4503-2257-7. DOI: 10.1145/2600428.2609615. 53, 57

G. Ottaviano, N. Tonellotto, and R. Venturini. Optimal space-time tradeoffs for inverted indexes. In *Proceedings of the 8th ACM International Conference on Web Search and Data Mining*, pages 47–56, New York, NY, USA, 2015. ISBN 978-1-4503-3317-7. DOI: 10.1145/2684822.2685297. 53, 57

I. Ounis, G. Amati, V. Plachouras, B. He, C. Macdonald, and C. Lioma. Terrier: A high performance and scalable information retrieval platform. In *Proceedings of ACM SIGIR'06 Workshop on Open Source Information Retrieval*, 2006. 78, 86

R. Ozcan, I. S. Altingovde, and O. Ulusoy. Static query result caching revisited. In *Proceedings of the 17th International Conference on World Wide Web*, pages 1169–1170, New York, NY, USA, 2008. ACM. ISBN 978-1-60558-085-2. DOI: 10.1145/1367497.1367710. 83, 88

R. Ozcan, I. S. Altingovde, B. B. Cambazoglu, F. P. Junqueira, and O. Ulusoy. A five-level static cache architecture for web search engines. *Inf. Process. Manage.*, 48(5):828–840, 2012. ISSN 0306-4573. DOI: 10.1016/j.ipm.2010.12.007. 84, 88

R. Ozcan, I. S. Altingovde, B. B. Cambazoglu, and O. Ulusoy. Second chance: A hybrid approach for dynamic result caching and prefetching in search engines. *ACM Trans. Web*, 8(1):3:1–3:22, December 2013. ISSN 1559-1131. DOI: 10.1145/2536777. 84, 88

S. Pandey and C. Olston. User-centric web crawling. In *Proceedings of the 14th International Conference on World Wide Web*, pages 401–411, New York, NY, USA, 2005. ACM. ISBN 1-59593-046-9. DOI: 10.1145/1060745.1060805. 25, 28

S. Pandey and C. Olston. Crawl ordering by search impact. In *Proceedings of the 1st International Conference on Web Search and Web Data Mining*, pages 3–14, New York, NY, USA, 2008. ACM. ISBN 978-1-59593-927-9. DOI: 10.1145/1341531.1341535. 25, 28

M. Persin, J. Zobel, and R. Sacks-Davis. Filtered document retrieval with frequency-sorted indexes. *J. Am. Soc. Inf. Sci.*, 47(10):749–764, 1996. ISSN 0002-8231. DOI: 10.1002/(SICI)1097-4571(199610)47:10%3C749::AID-ASI3%3E3.3.CO;2-U. 79, 87

D. Puppin, F. Silvestri, R. Perego, and R. Baeza-Yates. Tuning the capacity of search engines: Load-driven routing and incremental caching to reduce and balance the load. *ACM Trans. Inform. Syst.*, 28(2):1–36, 2010. ISSN 1046-8188. DOI: 10.1145/1740592.1740593. 82, 85, 88, 89

K. Radinsky and P. N. Bennett. Predicting content change on the Web. In *Proceedings of the 6th ACM International Conference on Web Search and Data Mining*, pages 415–424, New York, NY, USA, 2013. ISBN 978-1-4503-1869-3. DOI: 10.1145/2433396.2433448. 25, 28

S. Raghavan and H. Garcia-Molina. Crawling the hidden web. In *Proceedings of the 27th International Conference on Very Large Data Bases*, pages 129–138, San Francisco, CA, USA, 2001. Morgan Kaufmann Publishers Inc. ISBN 1-55860-804-4. 26, 29

B. Ribeiro-Neto, E. S. Moura, M. S. Neubert, and N. Ziviani. Efficient distributed algorithms to build inverted files. In *Proceedings of the 22nd Annual International ACM SIGIR Conference on Research and Development in Information Retrieval*, pages 105–112, New York, NY, USA, 1999. ISBN 1-58113-096-1. DOI: 10.1145/312624.312663. 54, 57

B. A. Ribeiro-Neto and R. A. Barbosa. Query performance for tightly coupled distributed digital libraries. In *Proceedings of the 3rd ACM conference on Digital Libraries*, pages 182–190, New York, NY, USA, 1998. ISBN 0-89791-965-3. DOI: 10.1145/276675.276695. 81, 87

B. A. Ribeiro-Neto, J. P. Kitajima, G. Navarro, C. R. G. Sant'Ana, and N. Ziviani. Parallel generation of inverted files for distributed text collections. In *Proceedings of the 18th International Conference of the Chilean Computer Science Society*, pages 149–157, Washington, DC, USA, 1998. IEEE Computer Society. ISBN 0-8186-8616-2. 54, 57

R. Rice and J. Plaunt. Adaptive variable-length coding for efficient compression of spacecraft television data. *Commun. Technol., IEEE Trans.*, 19(6):889–897, 1971. ISSN 0018-9332. DOI: 10.1109/TCOM.1971.1090789. 53, 56

K. M. Risvik, Y. Aasheim, and M. Lidal. Multi-tier architecture for web search engines. In *Proceedings of the 1st Conference on Latin American Web Congress*, page 132, Washington, DC, USA, 2003. IEEE Computer Society. ISBN 0-7695-2058-8. 83, 88

K. M. Risvik, T. Chilimbi, H. Tan, K. Kalyanaraman, and C. Anderson. Maguro, a system for indexing and searching over very large text collections. In *Proceedings of the 6th ACM International Conference on Web Search and Data Mining*, pages 727–736, New York, NY, USA, 2013. ISBN 978-1-4503-1869-3. DOI: 10.1145/2433396.2433486. 50, 56, 78, 86

P. C. Saraiva, E. S. de Moura, N. Ziviani, W. Meira, R. Fonseca, and B. Riberio-Neto. Rank-preserving two-level caching for scalable search engines. In *Proceedings of the 24th Annual International ACM SIGIR Conference on Research and Development in Information Retrieval*, pages 51–58, New York, NY, USA, 2001. ISBN 1-58113-331-6. DOI: 10.1145/383952.383959. 83, 84, 88

C. Sarigiannis, V. Plachouras, and R. Baeza-Yates. A study of the impact of index updates on distributed query processing for web search. In *Proceedings of the 31th European Conference on IR Research on Advances in Information Retrieval*, pages 595–602, Berlin, Heidelberg, 2009. Springer-Verlag. ISBN 978-3-642-00957-0. DOI: 10.1007/978-3-642-00958-7_55. 84, 89

F. B. Sazoglu, B. B. Cambazoglu, R. Ozcan, I. S. Altingovde, and O. Ulusoy. A financial cost metric for result caching. In *Proceedings of the 36th International ACM SIGIR Conference on Research and Development in Information Retrieval*, pages 873–876, New York, NY, USA, 2013. ISBN 978-1-4503-2034-4. DOI: 10.1145/2484028.2484182. 83, 88

U. Schonfeld and N. Shivakumar. Sitemaps: Above and beyond the crawl of duty. In *Proceedings of the 18th International Conference on World Wide Web*, pages 991–1000, New York, NY, USA, 2009. ACM. ISBN 978-1-60558-487-4. DOI: 10.1145/1526709.1526842. 26, 29

D. Shan, S. Ding, J. He, H. Yan, and X. Li. Optimized top-k processing with global page scores on block-max indexes. In *Proceedings of the 5th ACM International Conference on Web Search and Data Mining*, pages 423–432, New York, NY, USA, 2012. ISBN 978-1-4503-0747-5. DOI: 10.1145/2124295.2124346. 80, 87

W.-Y. Shieh and C.-P. Chung. A statistics-based approach to incrementally update inverted files. *Inf. Process. Manage.*, 41(2):275–288, 2005. ISSN 0306-4573. DOI: 10.1016/j.ipm.2003.10.004. 54, 57

W.-Y. Shieh, T.-F. Chen, J. J.-J. Shann, and C.-P. Chung. Inverted file compression through document identifier reassignment. *Inf. Process. Manage.*, 39(1):117–131, 2003. ISSN 0306-4573. DOI: 10.1016/S0306-4573(02)00020-1. 53, 57

V. Shkapenyuk and T. Suel. Design and implementation of a high-performance distributed web crawler. In *Proceedings of the 18th International Conference on Data Engineering*, pages 357–368, Washington, DC, USA, 2002. IEEE Computer Society. DOI: 10.1109/ICDE.2002.994750. 23, 28

L. Si and J. Callan. Relevant document distribution estimation method for resource selection. In *Proceedings of the 26th Annual International ACM SIGIR Conference on Research and Development in Information Retrieval*, pages 298–305, New York, NY, USA, 2003. ISBN 1-58113-646-3. DOI: 10.1145/860435.860490. 81, 82, 88

F. Silvestri. Sorting out the document identifier assignment problem. In G. Amati, C. Carpineto, and G. Romano, editors, *Advances in Information Retrieval*, volume 4425 of *Lecture Notes in Computer Science*, pages 101–112. Springer Berlin / Heidelberg, 2007. DOI: 10.1007/978-3-540-71496-5_12. 53, 57

F. Silvestri and R. Venturini. Vsencoding: Efficient coding and fast decoding of integer lists via dynamic programming. In *Proceedings of the 19th ACM International Conference on Information and Knowledge Management*, pages 1219–1228, New York, NY, USA, 2010. ISBN 978-1-4503-0099-5. DOI: 10.1145/1871437.1871592. 53, 56

F. Silvestri, S. Orlando, and R. Perego. Assigning identifiers to documents to enhance the clustering property of fulltext indexes. In *Proceedings of the 27th Annual International ACM SIGIR Conference on Research and Development in Information Retrieval*, pages 305–312, New York, NY, USA, 2004. ISBN 1-58113-881-4. DOI: 10.1145/1008992.1009046. 53, 57

A. Singh, M. Srivatsa, L. Liu, and T. Miller. Apoidea: A decentralized peer-to-peer architecture for crawling the World Wide Web. In Jamie Callan, Fabio Crestani, and Mark Sanderson, editors, *Distributed Multimedia Information Retrieval*, volume 2924 of *Lecture Notes in Computer Science*, pages 126–142. Springer Berlin Heidelberg, 2004. ISBN 978-3-540-20875-4. DOI: 10.1007/978-3-540-24610-7_10. 26, 28

G. Skobeltsyn, F. Junqueira, V. Plachouras, and R. Baeza-Yates. ResIn: a combination of results caching and index pruning for high-performance web search engines. In *Proceedings of the 31st Annual International ACM SIGIR Conference on Research and Development in Information Retrieval*, pages 131–138, New York, NY, USA, 2008. ISBN 978-1-60558-164-4. DOI: 10.1145/1390334.1390359. 85, 89

S. Sood and D. Loguinov. Probabilistic near-duplicate detection using simhash. In *Proceedings of the 20th ACM International Conference on Information and Knowledge Management*, pages 1117–1126, New York, NY, USA, 2011. ISBN 978-1-4503-0717-8. DOI: 10.1145/2063576.2063737. 52, 56

N. Spirin and J. Han. Survey on web spam detection: Principles and algorithms. *SIGKDD Explor. Newsl.*, 13(2):50–64, 2012. ISSN 1931-0145. DOI: 10.1145/2207243.2207252. 52, 56

A. A. Stepanov, A. R. Gangolli, D. E. Rose, R. J. Ernst, and P. S. Oberoi. SIMD-based decoding of posting lists. In *Proceedings of the 20th ACM International Conference on Information and Knowledge Management*, pages 317–326, New York, NY, USA, 2011. ISBN 978-1-4503-0717-8. DOI: 10.1145/2063576.2063627. 53, 56

T. Strohman, D. Metzler, H. Turtle, and W. B. Croft. Indri: A language model-based search engine for complex queries. In *Proceedings of the International Conference on Intelligence Analysis*, 2005a. 78, 86

T. Strohman and W. B. Croft. Efficient document retrieval in main memory. In *Proceedings of the 30th Annual International ACM SIGIR Conference on Research and Development in Information Retrieval*, pages 175–182, New York, NY, USA, 2007. ISBN 978-1-59593-597-7. DOI: 10.1145/1277741.1277774. 79, 87

T. Strohman, H. Turtle, and W. B. Croft. Optimization strategies for complex queries. In *Proceedings of the 28th Annual International ACM SIGIR Conference on Research and Development in Information Retrieval*, pages 219–225, New York, NY, USA, 2005b. ISBN 1-59593-034-5. DOI: 10.1145/1076034.1076074. 79, 87

T. Suel, C. Mathur, J.-W. Wu, J. Zhang, A. Delis, M. Kharrazi, X. Long, and K. Shanmugasundaram. ODISSEA: A peer-to-peer architecture for scalable web search and information retrieval. In V. Christophides and J. Freire, editors, *WebDB*, pages 67–72, 2003. 26, 28

Q. Tan and P. Mitra. Clustering-based incremental web crawling. *ACM Trans. Inf. Syst.*, 28(4): 17:1–17:27, 2010. ISSN 1046-8188. DOI: 10.1145/1852102.1852103. 25, 28

X. Tang, X. Jin, and T. Yang. Cache-conscious runtime optimization for ranking ensembles. In *Proceedings of the 37th International ACM SIGIR Conference on Research and Development in Information Retrieval*, pages 1123–1126, New York, NY, USA, 2014. ISBN 978-1-4503-2257-7. DOI: 10.1145/2600428.2609525. 85, 89

A. Teymorian, O. Frieder, and M. A. Maloof. Rank-energy selective query forwarding for distributed search systems. In *Proceedings of the 22nd ACM International Conference on Information and Knowledge Management*, pages 389–398, New York, NY, USA, 2013. ISBN 978-1-4503-2263-8. DOI: 10.1145/2505515.2505710. 84, 85, 89

M. Theobald, J. Siddharth, and A. Paepcke. SpotSigs: Robust and efficient near duplicate detection in large web collections. In *Proceedings of the 31st Annual International ACM SIGIR Conference on Research and Development in Information Retrieval*, pages 563–570, New York, NY, USA, 2008. ISBN 978-1-60558-164-4. DOI: 10.1145/1390334.1390431. 52, 56

S. L. Thota and B. Carterette. Within-document term-based index pruning with statistical hypothesis testing. In *Proceedings of the 33rd European Conference on Advances in Information Retrieval*, pages 543–554, Berlin, Heidelberg, 2011. Springer-Verlag. ISBN 978-3-642-20160-8. 82, 83, 88

A. Tomasic and H. Garcia-Molina. Caching and database scaling in distributed shared-nothing information retrieval systems. *ACM SIGMOD Record*, 22(2):129–138, 1993a. ISSN 0163-5808. DOI: 10.1145/170036.170063. 83, 88

A. Tomasic and H. Garcia-Molina. Performance of inverted indices in shared-nothing distributed text document informatioon retrieval systems. In *Proceedings of the 2nd International*

Conference on Parallel and Distributed Information Systems, pages 8–17, Los Alamitos, CA, USA, 1993b. IEEE Computer Society Press. ISBN 0-8186-330-1. 81, 87

A. Tomasic, H. García-Molina, and K. Shoens. Incremental updates of inverted lists for text document retrieval. In *Proceedings of the 1994 ACM SIGMOD International Conference on Management of Data*, pages 289–300, New York, NY, USA, 1994. ISBN 0-89791-639-5. DOI: 10.1145/191839.191896. 55, 57

N. Tonellotto, C. Macdonald, and I. Ounis. Effect of different docid orderings on dynamic pruning retrieval strategies. In *Proceedings of the 34th International ACM SIGIR Conference on Research and Development in Information Retrieval*, pages 1179–1180, New York, NY, USA, 2011. ISBN 978-1-4503-0757-4. DOI: 10.1145/2009916.2010108. 85, 89

N. Tonellotto, C. Macdonald, and I. Ounis. Efficient and effective retrieval using selective pruning. In *Proceedings of the 6th ACM International Conference on Web Search and Data Mining*, pages 63–72, New York, NY, USA, 2013. ISBN 978-1-4503-1869-3. DOI: 10.1145/2433396.2433407. 80, 87

G. Tran, A. Turk, B. B. Cambazoglu, and W. Nejdl. A random walk model for optimization of search impact in web frontier ranking. In *Proceedings of the 38th International ACM SIGIR Conference on Research and Development in Information Retrieval*, New York, NY, USA, 2015. 25, 28

Y. Tsegay, A. Turpin, and J. Zobel. Dynamic index pruning for effective caching. In *Proceedings of the 16th ACM Conference on Information and Knowledge Management*, pages 987–990, New York, NY, USA, 2007. ISBN 978-1-59593-803-9. DOI: 10.1145/1321440.1321592. 85, 89

A. Turpin, Y. Tsegay, D. Hawking, and H. E. Williams. Fast generation of result snippets in web search. In *Proceedings of the 30th Annual International ACM SIGIR Conference on Research and Development in Information Retrieval*, pages 127–134, New York, NY, USA, 2007. ISBN 978-1-59593-597-7. DOI: 10.1145/1277741.1277766. 84, 88

H. Turtle and J. Flood. Query evaluation: Strategies and optimizations. *Inf. Process. Manage.*, 31 (6):831–850, 1995. ISSN 0306-4573. DOI: 10.1016/0306-4573(95)00020-H. 79, 87

T. Urvoy, E. Chauveau, P. Filoche, and T. Lavergne. Tracking web spam with HTML style similarities. *ACM Trans. Web*, 2(1):3:1–3:28, 2008. ISSN 1559-1131. DOI: 10.1145/1326561.1326564. 52, 56

S. Vigna. Quasi-succinct indices. In *Proceedings of the 6th ACM International Conference on Web Search and Data Mining*, pages 83–92, New York, NY, USA, 2013. ISBN 978-1-4503-1869-3. DOI: 10.1145/2433396.2433409. 53, 57

A. G. Vural, B. B. Cambazoglu, and P. Senkul. Sentiment-focused web crawling. In *Proceedings of the 21st ACM International Conference on Information and Knowledge Management*, pages 2020–2024, New York, NY, USA, 2012. ISBN 978-1-4503-1156-4. DOI: 10.1145/2396761.2398564. 26, 29

J. Wang, E. Lo, M. L. Yiu, J. Tong, G. Wang, and X. Liu. The impact of solid state drive on search engine cache management. In *Proceedings of the 36th International ACM SIGIR Conference on Research and Development in Information Retrieval*, pages 693–702, New York, NY, USA, 2013. ISBN 978-1-4503-2034-4. DOI: 10.1145/2484028.2484046. 84, 88, 92

L. Wang, J. Lin, and D. Metzler. Learning to efficiently rank. In *Proceedings of the 33rd International ACM SIGIR Conference on Research and Development in Information Retrieval*, pages 138–145, New York, NY, USA, 2010. ISBN 978-1-4503-0153-4. DOI: 10.1145/1835449.1835475. 85, 89

L. Wang, J. Lin, and D. Metzler. A cascade ranking model for efficient ranked retrieval. In *Proceedings of the 34th International ACM SIGIR Conference on Research and Development in Information Retrieval*, pages 105–114, New York, NY, USA, 2011. ISBN 978-1-4503-0757-4. DOI: 10.1145/2009916.2009934. 85, 89

S. Webb, J. Caverlee, and C. Pu. Predicting web spam with HTTP session information. In *Proceedings of the 17th ACM Conference on Information and Knowledge Management*, pages 339–348, New York, NY, USA, 2008. ISBN 978-1-59593-991-3. DOI: 10.1145/1458082.1458129. 26, 29

I. H. Witten, A. Moffat, and T. C. Bell. *Managing Gigabytes: Compressing and Indexing Documents and Images*. Morgan Kaufmann Publishers Inc., San Francisco, CA, USA, 2nd edition, 1999. ISBN 1-55860-570-3. 55, 56, 92

J. L. Wolf, M. S. Squillante, P. S. Yu, J. Sethuraman, and L. Ozsen. Optimal crawling strategies for web search engines. In *Proceedings of the 11th International Conference on World Wide Web*, pages 136–147, New York, NY, USA, 2002. ACM. ISBN 1-58113-449-5. DOI: 10.1145/511446.511465. 25, 28

B. Wu and B. D. Davison. Detecting semantic cloaking on the Web. In *Proceedings of the 15th International Conference on World Wide Web*, pages 819–828, New York, NY, USA, 2006. ISBN 1-59593-323-9. DOI: 10.1145/1135777.1135901. 53, 56

B. Wu, V. Goel, and B. D. Davison. Topical TrustRank: Using topicality to combat web spam. In *Proceedings of the 15th International Conference on World Wide Web*, pages 63–72, New York, NY, USA, 2006. ACM. ISBN 1-59593-323-9. DOI: 10.1145/1135777.1135792. 52, 53, 56

W. Xi, O. Sornil, M. Luo, and E. A. Fox. Hybrid partition inverted files: experimental validation. In M. Agosti and C. Thanos, editors, *Research and Advanced Technology for Digital Libraries*, volume 2458 of *Lecture Notes in Computer Science*, pages 422–431. Springer Berlin Heidelberg, 2002. ISBN 978-3-540-44178-6. DOI: 10.1007/3-540-45747-X_31. 81, 87

H. Yan, S. Ding, and T. Suel. Compressing term positions in web indexes. In *Proceedings of the 32nd International ACM SIGIR Conference on Research and Development in Information Retrieval*, pages 147–154, New York, NY, USA, 2009a. ISBN 978-1-60558-483-6. DOI: 10.1145/1571941.1571969. 53, 57

H. Yan, S. Ding, and T. Suel. Inverted index compression and query processing with optimized document ordering. In *Proceedings of the 18th International Conference on World Wide Web*, pages 401–410, New York, NY, USA, 2009b. ACM. ISBN 978-1-60558-487-4. DOI: 10.1145/1526709.1526764. 53, 56, 57

F. Zhang, S. Shi, H. Yan, and J.-R. Wen. Revisiting globally sorted indexes for efficient document retrieval. In *Proceedings of the 3rd ACM International Conference on Web Search and Data Mining*, pages 371–380, New York, NY, USA, 2010. ISBN 978-1-60558-889-6. DOI: 10.1145/1718487.1718534. 80, 87

J. Zhang and T. Suel. Optimized inverted list assignment in distributed search engine architectures. In *IEEE International Parallel and Distributed Processing Symposium*, pages 1–10, 2007. DOI: 10.1109/IPDPS.2007.370231. 81, 87

J. Zhang, X. Long, and T. Suel. Performance of compressed inverted list caching in search engines. In *Proceedings of the 17th International Conference on World Wide Web*, pages 387–396, New York, NY, USA, 2008. ACM. ISBN 978-1-60558-085-2. DOI: 10.1145/1367497.1367550. 53, 56, 85, 89

J. Zobel and A. Moffat. Inverted files for text search engines. *ACM Comput. Surv.*, 38(2), 2006. ISSN 0360-0300. DOI: 10.1145/1132956.1132959. 55, 56

J. Zobel, A. Moffat, and K. Ramamohanarao. Inverted files versus signature files for text indexing. *ACM Trans. Database Syst.*, 23(4):453–490, 1998. ISSN 0362-5915. DOI: 10.1145/296854.277632. 50, 56

M. Zukowski, S. Heman, N. Nes, and P. Boncz. Super-scalar RAM-CPU cache compression. In *Proceedings of the 22nd International Conference on Data Engineering*, pages 59–, Washington, DC, USA, 2006. IEEE Computer Society. ISBN 0-7695-2570-9. DOI: 10.1109/ICDE.2006.150. 53, 56

Authors' Biographies

B. BARLA CAMBAZOGLU

B. Barla Cambazoglu received his B.S., M.S., and Ph.D. degrees, all in computer engineering, from the Computer Engineering Department of Bilkent University in 1997, 2000, and 2006, respectively. After getting his Ph.D. degree, he worked as a postdoctoral researcher in Bilkent University for a short period of time. In 2006, he joined the Biomedical Informatics Department of the Ohio State University as a postdoctoral researcher. In 2008, he joined Yahoo Labs as a postdoctoral researcher. He received research scientist and senior research scientist positions at the same institution, in 2010 and 2012, respectively. Between 2013 and 2015, he was a senior manager, heading the web retrieval group in Yahoo Labs Barcelona. His main research interests are distributed information retrieval and web search efficiency. In 2010, 2011, 2014, and 2015, he co-organized the LSDS-IR workshop. He was the proceedings chair for WSDM'09 and the poster and proceedings chairs for ECIR'12. He served as an area chair in SIGIR'13 and SIGIR'14. He regularly serves on the program committees of SIGIR, WWW, and KDD conferences. He has many papers published in prestigious journals including IEEE TPDS, JPDC, JASIST, Inf. Syst., ACM TWEB, and IP&M, as well as papers and tutorials presented at top-tier conferences, such as SIGIR, CIKM, WSDM, WWW, and KDD.

RICARDO BAEZA-YATES

Ricardo Baeza-Yates has been VP of Research and Chief Research Scientist at Yahoo Labs, based in Sunnyvale, California, since August 2014. Before that, he founded and led the labs in Barcelona and Santiago de Chile from 2006–2015. Between 2008 and 2012 he also oversaw the Haifa lab. In addition, he is also a part-time Professor at the Department of Information and Communication Technologies of the Universitat Pompeu Fabra, in Barcelona, Spain, where in 2005 he was an ICREA research professor. Until 2004 he was a Professor, and before that founder and Director, of the Center for Web Research at the Department of Computing Science of the University of Chile (from where he is currently on a leave of absence). In 1989, he obtained a Ph.D. in computer science from the University of Waterloo, Canada. Before that, he obtained two master degrees (M.Sc. CS & M.Eng. EE) and an electronic engineering degree from the University of Chile in Santiago. He is co-author of the best-seller Modern Information Retrieval textbook, published in 1999 by Addison-Wesley, with a second enlarged edition in 2011, that won the ASIST 2012 Book of the Year award. He is also co-author of the 2nd edition of the Handbook of Algorithms and Data Structures, Addison-Wesley, 1991 and co-editor of Information Retrieval: Algorithms and Data Structures, Prentice-Hall, 1992. In addition, he is the author or co-author of more than 500 other publications. From 2002-2004, he was elected to the board of governors of the IEEE Computer Society and in 2012 he was elected for the ACM Council. He received the Organization of American States award for young researchers in exact sciences (1993), the Graham Medal for innovation in computing given by the University of Waterloo to distinguished ex-alumni (2007), the CLEI Latin American distinction for contributions to CS in the region (2009), and the National Award of the Chilean Association of Engineers (2010), among other distinctions. In 2003 he was the first computer scientist to be elected to the Chilean Academy of Sciences and since 2010 has been a founding member of the Chilean Academy of Engineering. In 2009 he was named ACM Fellow and in 2011 IEEE Fellow.

Printed in the United States
by Baker & Taylor Publisher Services